Ribbon of Sand

A Chapel Hill Book

Ribbon of Sand

The
Amazing
Convergence
of the
Ocean
& the
Outer
Banks

John Alexander
& James Lazell

The University of North Carolina Press
Chapel Hill & London

Originally published in 1992 by Algonquin Books of
Chapel Hill. Paperback edition published in 2000 by the
University of North Carolina Press by arrangement with
Algonquin Books of Chapel Hill.

Manufactured in the United States of America
The paper in this book meets the guidelines for permanence
and durability of the Committee on Production Guidelines for
Book Longevity of the Council on Library Resources.

Library of Congress Cataloging-in-Publication Data
Alexander, John, 1945 Sept. 14–
Ribbon of sand: the amazing convergence of the ocean and
the Outer Banks / by John Alexander and James Lazell.
    p.   cm.
Originally published: Chapel Hill: Algonquin Books, 1992.
Includes bibliograp'   eferences and index.
ISBN 0-8078-4874   k. paper)
1. Outer Banks (N.C.)   .   2. Outer Banks (N.C.)—
Geography.   3. Ecology   Carolina—Outer Banks.
I. Lazell, James D.   II. T
F262.O96 A44   2000   9   21   99-057677

04 03 02 01 00   5 4 3 2 1

Contents

Maps & Diagrams

Preface

O N A GEOLOGICAL SCALE, were the Outer Banks left to themselves, not a lot would have happened there in the eight years since this book was first published. Seas rise and fall; storms blow in and add their touches to the landscape; sand continues its unceasing movement with the wind and tide.

As any visitor knows, however, the Outer Banks most definitely are not left to themselves, and in human terms a good deal has happened there since 1992. Some of the interactions between the Outer Banks and their human inhabitants—matters discussed in the last chapter, "Convergences"—have altered significantly.

Mobil Oil Corporation's proposal to drill for natural gas and oil at a site they call Manteo Prospect, off the coast of Cape Hatteras, is still stalled in federal court. In September 1997, however, Chevron Oil announced a separate plan to drill at a spot just 9,000 feet from Manteo Prospect beginning in the spring of 2000. The Division of Coastal Management has set aside $367,000 in its budget for review of Chevron's application.

Government support for shoring up Oregon Inlet has increased, and both Senator Jesse Helms and Governor Jim Hunt endorsed the idea in 1996. In November 1998, legislation was passed giving the Wanchese Seafood Industrial Park Authority responsibility for the inlet. Although no formal plans have been announced, it is expected that the Wanchese Authority will attempt to reclaim the land and build the proposed jetties.

In 1995, in their monumental study, *Reptiles of North Carolina* (University of North Carolina Press), Drs. William Palmer and Alvin Braswell of the North Carolina State Museum of Natural Sciences did not regard the *sticticeps* kingsnake as a valid subspecies. They amalgamated all Outer Banks kingsnakes—ordinary mainland types and intergrades from north of The Line with real *sticticeps* from south of The Line—and made a composite that is, indeed, not distinct. The snake in their figure 91 is not a *sticticeps* and is from north of The Line. They used a method of counting scales different from the one we used and thus got consistently higher counts, both on and off the Banks. No biologists' views, however, alter the distinctive character of the bizarre kingsnakes found south of The Line down to Cape Lookout.

In June and July of 1999, after protracted legal battles, the Cape Hatteras Lighthouse was moved by the International Chimney Corporation of New York. The move took twenty-two days, during which time an estimated ten thousand people a day came to view the process. The lighthouse will be formally reopened to visitors on Memorial Day, 2000.

Off the coast of Atlantic Beach, in November 1996, scientists discovered what they believe to be the wreck of Blackbeard's flagship, the *Queen Anne's Revenge*. Excavation of the wreck continues today, but so far no treasure of gold has been raised from the water. Many of the plainer treasures of history—pewter, lead, brass, iron—have been brought to the surface, though, providing insights into Blackbeard's legend and the times when he lived. As the Outer Banks prepares to celebrate the hundredth anniversary of Orville and Wilbur Wright's first flight at Kitty Hawk, perhaps this resurfacing can serve as another reminder that our history, and our future, remains tied to the land and the sea.

December 1999

Acknowledgments

THIS IS A BOOK of connections. Its central character, the connector, the Outer Banks, fringes the eastern edge of our continent. The stories we present here bear the philosophical imprint of John Muir, who saw that ecologically everything is linked to everything else, and of James Burke, author of *Connections,* who wrote: "The reason why each event took place where and when it did is a fascinating mixture of accident, climatic change, genius, craftsmanship, careful observation, ambition... and a hundred other factors."

The book has its origins at the University of the South in Sewanee, Tennessee, in 1957, when the authors first began observing and collecting animals together. That friendship and collaboration have spanned more than thirty years, during which we were both drawn to the sands of the Outer Banks in search of an odd kingsnake. Our first joint visit to the Outer Banks occurred in 1972. But it was not until more than fifteen years later that we decided to write this volume.

For their generous support of the research and writing of this book, we are indebted to the Hillsdale Fund and the Mary Norris Preyer Fund of Greensboro, North Carolina; the Prospect Hill Foundation of New York; The Conservation Agency of Rhode Island; the National Park Service; Algonquin Books of Chapel Hill; and our editor, Louis D. Rubin, Jr. And to all who assisted along the way—catching snakes, cleaning fish, rowing boats, and sweeping sand—our gratitude.

The living inhabitation of the world . . . , the
spiritual power of the air, the rocks, the waters,
to be in the midst of it, and rejoice and wonder at
it, and help it if I could, . . . this was the essential
love of Nature in me, this the root of all that I
have usefully become, and the light of all that
I have rightly learned.

John Ruskin, 1899

                                               Light
flows from the water from sands   islands of this horizon
The sea comes toward me across the sea.   The sand
moves over the sand in waves
between the guardians of this landscape
the great commemorative statue on one hand
      —the first flight of man, outside of dream,
       seen as stone wing and stainless steel—
and at the other hand
      banded black-and-white, climbing
the spiral lighthouse.

Muriel Rukeyser, 1980

Ribbon of Sand

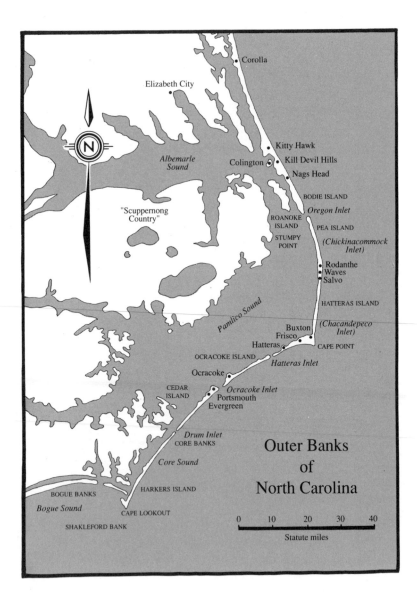

Corolla

Elizabeth City

Kitty Hawk

Kill Devil Hills

Colington

Nags Head

Albemarle Sound

BODIE ISLAND

Oregon Inlet

"Scuppernong Country"

ROANOKE ISLAND

PEA ISLAND

STUMPY POINT

(Chickinacommock Inlet)

Rodanthe

Waves

Salvo

Pamlico Sound

HATTERAS ISLAND

(Chacandepeco Inlet)

Buxton

Frisco

Hatteras

CAPE POINT

OCRACOKE ISLAND

Hatteras Inlet

Ocracoke

CEDAR ISLAND

Ocracoke Inlet

Portsmouth

Evergreen

Drum Inlet

CORE BANKS

Core Sound

HARKERS ISLAND

BOGUE BANKS

Bogue Sound

CAPE LOOKOUT

SHAKLEFORD BANK

**Outer Banks
of
North Carolina**

0   10   20   30   40

Statute miles

Chapter One

Sand

And everyone that heareth these sayings of mine,
and doeth them not, shall be likened unto a
foolish man, which buildeth his house upon
the sand: And the rain descended, and
the floods came, and the winds blew
and beat upon that house; and it
fell: and great was the fall of it.
Matthew 7: 26–27

But the sand! The sand is the greatest thing
in Kitty Hawk, and soon will be the only thing. . . .
The sea has washed and the wind blown millions
and millions of loads of sand up in heaps along
the coast, completely covering houses and forest.
Orville Wright, 1900

J UNE NIGHT: Cape Hatteras, North Carolina. The Cape Hatteras Lighthouse, beacon to sailors, blinks seaward. With its black-and-white candy-stripe pattern, the lighthouse has proudly symbolized both the romance and terror of the Atlantic for more than a century. The nation's oldest and tallest brick lighthouse, it has stood watch over more shipwrecks than any other. But on this night it symbolizes something else: the churlish power of the sea and sand together to humble the grandest human labors, to claim them as surely as it claims a child's sand castle or ghost crab's burrow.

As waves lap ever closer to the lighthouse's timber-and-granite base, the brick structure is losing its precarious perch. In due time, according to a schedule that only the ocean keeps, the sand around its footing will begin to wash away. Barring a cumbersome, expensive, engineers' scheme to rescue the structure by moving it landward, this stately monument—littoral leaning tower—will surrender to the sea's indifferent embrace.

On this warm June night, however, the lighthouse is not the center of attention: far from it. Man's futile struggle to spare the edifice is only the most conspicuous chapter of a longer epic that has preceded human settlement on the Outer Banks, and that is certain to outlast it. It is an epic of sand, wind, and sea that defines and directs everything that dwells on the Banks, like some epochal operatic production: *Aïda* without the elephants. The difference is that the Banks' nonhuman inhabitants must adapt to their environment, or they will die; humans, unthreatened, try to tame it. But the sand, great equalizer, ultimately tames all.

All along Cape Hatteras, a late spring northeaster shoots large, warm, restorative drops of rain. Above the clouds, hidden from the beach a mile north of the lighthouse, a full moon routs the darkness. This is no ordinary night, though the ritual that is about to unfold has played countless times before. On this night, a giant wave blankets the beach along a 100-meter stretch. It is the largest wave in the highest of the month's high-course tides. Pushed by the northeast wind (winds are named for the direction they come *from*), the wave is the largest of the last set at flood tide. Its pounding carries farthest up the beach into the dunes. It contains 4,000 metric tons of water and one-and-a-half million pounds of sand.

But water and sand are not the wave's only cargo. It hauls small animals of the surf zone—the little crustaceans called sand crabs, uncounted and unweighed: imponderable. Also buoyed in its roiling, hissing mass is one ponderous particle: 638 pounds of loggerhead sea turtle. Unlike the sand, the turtle has done everything she can to help the wave move her forward. Her great sculling flippers have scooped into the moving water. Now, with all its curling power, the wave shoves the female forward, disgorging her like excess cargo, wet and not far from the surf. Wet, yes, but as dry as she has been for nearly two years, since the last time she undertook the perilous journey from sea to land.

The peril is as palpable as the warm rain splashing on the turtle's barnacled shell—peril not so much to the turtle, though human predators still threaten her species' survival, but to her burden of round, white eggs. It is not enough that only a small number of the hatchlings gestating in those eggs will ever migrate back to sea. The odds of winning at poker or blackjack are much better.

The threat is more immediate, in the form of a ravenous mother raccoon and her two remaining kits. A third, the weakest, has died already. The raccoons have made their home in an old loblolly pine trunk in the woods west of the lighthouse. It is a spacious den, but these are hard times: too many raccoons. Real summer, just around the corner, will bring people in droves, and people will bring garbage— succulent, mouth-watering quantities of it. For now, there is little to forage, unless the mother raccoon's bad luck improves. Unless, for example, she digs for buried treasure—a clutch of turtle eggs.

THE OUTER BANKS are made of sand. True, there are firm peats, produced on the sound side, of roots and stems and mud. There are marls, formed of clay and shell and pebbles. There are loams developed from humus in the high hammocks and stately forests. And, in the sloughs and back lagoons, there is wondrous, deep, black ooze. All of these are, however, but thin layers and swirls of differently flavored and textured icing—pretty confections—applied to the foundation of sand. Yet for all its aggregate firmness, sand is never still, always dynamic. From the inherent instability of sand, nature has produced a series of the most tenacious and resilient ecological communities on earth.

Consider a grain of sand. Alone, it is no more than two millimeters in diameter, smaller than an ant's back. The faintest breath of wind can lift it and turn it endlessly on itself, like a circus tumbler. Rushing water, whether wave or river current, sends it flying.

Though sand occurs the world over—it is the ubiquitous mortar of earth's geological undulations—we associate it with deserts and oceans, where it is most abundant and visible. While we have learned that seemingly hostile desert sands teem with life, the desert's arid environment is largely inhospitable to humans, except for the occasional prospector, hermit, or sports car advertisement crew. Most of us prefer the beach's restorative powers, the squeak of bare toes digging into sand. (Why do feet squeak in sand? Why not merely crunch or trudge?) This gritty affinity, like that of pelicans to pilings, is only natural. The ocean is the watery womb from which our evolutionary ancestors slipped, slithered, or crawled. And when they poked forth their primitive limbs or fins or whatever appendage came first, it must have been into sand, pools of it, the same porous stuff in which we wiggle our toes and construct Arthurian castles.

(Some take it more seriously. Members of a cult claiming to mimic Indian ritual recently buried several initiates under eighteen inches of sand at North Carolina's Topsail Beach. Despite the use of snorkels for breathing, one of the recruits suffocated and died. The local prosecutor did not press charges against the cult's leaders, arguing that the poor soul underwent the burial voluntarily and in full knowledge of the risk. Is there a statute of limitations on crimes committed with sand?)

For the record, two types of sand particles find their way

into our blankets and coolers at a typical Atlantic Coast beach. *Siliceous* sand is made of oxides of silicon derived from pulverized rock. *Calcareous* sand comes from calcium-based, pulverized seashells. Apart from their different molecular composition, the two varieties of sand differ in shape. Siliceous sand grains come in chunks, approaching cubes or spheres; calcareous sand comes in flakes. The flakes are relatively flat and thin. In water or in air, they have a lot of lift. They carry farther.

As you might imagine, the composition of sand deposits along the Outer Banks depends upon the sand's source. Siliceous sand comes from the earth's commonest sort of rock. Vast amounts are ground up in the movement of glaciers, then released by melting during interglacial periods—like today.

When sea level was lower, rivers carried sand across the continental shelf and formed large deposits. These deposits, since covered and massaged by the sea, form parts of today's beaches. Now that sea level has risen again, most sand carried by rivers is deposited in bays and sounds, not on the beach itself. Only the finest sand is carried seaward. To some degree, the same old sand that forms the Outer Banks is rolled over itself again and again—more proof that there is nothing new under the sun.

Now scoop up a handful of sand. Though a cup of it, dry, weighs no more than a pound and a third, barely enough to fill an unsuspecting pant cuff, more grains are already accumulated there than could be comfortably counted in a summer afternoon, or a succession of afternoons. Few children have ever received a satisfactory answer to the perennial question: how many grains of sand does a beach contain? Which is presumably why children continue to ask it, unless it is to exasperate the inquiring minds of parents. One might as well ask how many mosquitos swarm the beach air on a warm, windless night: too many to care, or swat.

Since scientists presumably have more urgent things to occupy them, the quest for an answer may best be left to poets, who specialize in such imprecisions. But it is a fact nonetheless that the beach is made up of individual particles of sand, unimaginably vast quantities of them. If the mathematical concept of infinity has meaning at all, it is surely contained in the sand splayed under our feet, just as it can be imagined in the thick canopy of stars over our heads.

When the wind blows, as it often does in flat coastal areas, it pushes sand into dunes. When ocean storms buffet the coast, as they often do, waves move literally tons of sand. To be specific, a wave 250 yards long would move about 1½ million pounds, or nearly 24,000 cubic feet, of sand. The wave itself would weigh about 4,000 tons—about 140,000 cubic feet of water.

That enormous quantity, of course, is only a small cross-section of the sand-moving abilities of ocean waves and currents. There are no ninety-pound weaklings here. North of Hatteras, near Oregon Inlet, scientists estimate that between one-half and one million cubic meters of sand are transported each year—due south, toward Cape Hatteras. Nature's mass-transit sand system truly boggles the mind.

IT IS JUST such a hissing, roiling wave that has catapulted the female loggerhead turtle onto the beach at Cape Hatteras. At this time of year, June, along this stretch of coast, the sand is about 20 percent calcareous. Because of their lift, these flakes will increase in percentage of the total as the wave rushes upslope toward the dune crest. If the wave is big enough and pushed hard enough to carry beyond the dune crest to overwash the interdune swale, most of the sand it carries that far will be of the calcareous variety. That is important not only to the survival of the turtle and her cargo of eggs, but also to the dynamic process by which wind and waves move sand. It is, in its barest essentials, the story of that thin, shifting, ribbon of sand called the Outer Banks.

The raccoons, however, do not know this—do not know or care about the geology of the place. The sand is their dinner table, and the menu, while delectable to read, lacks support from the kitchen. The raccoons have set off early in the evening, well before high tide. All three are weak. The kits, weaned now, hunt in desultory fashion, but their mother's senses are tuned by desperation. They are well north of the narrow waist of beach where the turtle first sprawled. Had the raccoons encountered the turtle, they would have waited, impatiently, until she laid her eggs. Like ill-mannered guests at a boarding house, the raccoons would have gorged on those eggs, then holed up nearby and returned to feast until not a single egg remained.

The turtle's sweet eggs would surely have saved the raccoons, if only for a short while, and—figuring two clutches a summer every other year over a span of six years—would have cost the turtle more than 16 percent of her total lifetime reproductive effort.

But the raccoons do not yet sense the turtle's presence nearby. The determined female has taken an hour to heave herself up the beach, up a sea oat–crowned dune, to her chosen nest site near the top. She has consumed three hours in excavating her nest, back flippers alternating, scooping sand. In another hour she has deposited her eggs. Then, rheumy-eyed, ponderous, and coated with sand, she heaves herself back across the beach and slips into the warm surf. Her mission accomplished, she can now trade her lumbering gait for the swimmer's graceful strokes. If female loggerheads dream at such times, it must be of exhausting, enveloping, sustaining coverlets of sand.

Dawn. The weather has changed. The northeast wind has died, the rain has ceased, and the clouds have broken into lumps scudding eastward out to sea. The raccoons are more afraid of daylight than they are of another day without food. They curl beneath a driftwood stump in the dunes about a quarter mile north of the turtle's nest.

Anyone can see the turtle's clear, regular flipper bites in the sand, and the broad line where she dragged her shell up the incline and back down again. Her crawl has left a huge V across the beach. At the apex of the V, the high dune bears the flattened, smooth surface of her subcircular nest. The nest location is completely exposed and, had the raccoons wandered there, they could have smelled it. They could have found it as effortlessly as if they had stolen upon the turtle laying her eggs. This time, they were foraging just a little too far away.

By midmorning the rising tide is erasing the lower hieroglyphics of the turtle's crawl. The sun bursts through lingering clouds. The wind freshens now from the opposite direction, southwest, its normal quarter of origin at this latitude, more than thirty-five degrees above the equator. By noon the previous night's wet sand is drying rapidly. Waves have obliterated most of the turtle's signature, but the peak of the V and, of course, the nest itself—well above the reach of the high-course tide waves—are still exposed.

In the swale the new sand grains begin to move: the wind budges them. First a light flake of calcareous sand (a minute fragment of a Massachusetts clam shell), flips over. The sun and wind reverse last night's direction of water transfer; they reclaim the sand's moisture as vapor.

Like minuscule soldiers, the drying sand grains break and scurry up the dune's back face. First one, then two and three leap up and charge forward northeast, roughly the direction the wave had brought them. A gust of southwest wind in the westering sun brings forth a battalion. They scurry up the dune, zigzagging from one small obstacle to another, only to fall behind the cover of a twig, a wisp of oat straw, a stem. An army of sand grains is capturing the high ground, the mountain of dune.

The first sand grains to reach the dune's crest toboggan down the seaward face, freed from their climb. Soon a gust brings millions more over the edge, a cloud of conquering, calcareous sand grains flying and parachuting down. They land lightly, flip a time or two and hurtle over the turtle's nest. They fill the foxholes of her remnant tracks. By dusk this inert army has taken all. No trace of the turtle's crawl marks remain. The nest is completely buried. The turtle's rich aromas are no longer sniffable at the surface, by raccoons or any other predator. It is, after all, a new surface, randomly wind-sculpted. The sand-grain army has landed, has taken first the low ground, swarmed the high triumphantly, and encamped in its millions all over the eastern front. No raccoon will ever find that nest.

One 'coon kit never leaves the shelter of the driftwood the next morning. Within days his bones are picked clean by ghost crabs whose tunnels converged through the wet sand on the underside of the stump. The mother and remaining kit, now thirsty, too, head back south, passing within ten feet of the turtle's camouflaged nest. They will never know it.

The raccoons cut westward through the dune swales to the narrow, blacktop road. A plump ghost crab heavy with roe moved too slowly to avoid an oncoming automobile at dusk. She now becomes a lifesaving appetizer for the raccoons. Mother and kit do not peacefully divide this small meal; they fight over it. The smaller kit proves more agile. He wrests away the bulk of the crab's body and bolts into the sea oats flanking the road. His mother, intent on legs and gobbets, never sees the car. Absorbed in his feast, the kit pays scant heed to the brightening, then dimming, lights as the car speeds onward. Finished but unsated, he crosses the road and works the marsh edge southward, alone.

A water snake is also casing the marsh edge, moving eastward along a little lead or "creek" that becomes progressively brackish as it winds inland. Once that little creek was the channel of a great inlet, cutting the Banks from sea to sound. But that was in the year A.D. 1585, by human reckoning. Sand and storm and wind have reduced it to a quiet creek. Rainwater has rendered it brackish.

Desperately thirsty, the young raccoon tastes the water as he moves. The freshening gradient spurs him on. The tide is low now, and he can almost trot along in ankle-deep water on the creek's mud floor. The snake also searches for water fresh enough to drink and almost finds it. It reacts to the movement of the advancing raccoon by lunging into the deeper water, swimming in sinuous curves, lashing the creek's surface to escape. An hour later the incoming tide would have been high enough to force the raccoon to swim. He would not have been equipped to swim as fast as the snake. But now the bottom still supports the wade-running raccoon. The water snake's hour has come.

The raccoon forges up the creek, the writhing but mostly dead water snake gripped in his teeth. The water has become fresh enough for the raccoon to drink; the sumptuous meal rekindles his zeal for the hunt. Using his newfound strength, he crushes the shells of ribbed mussels. Long before dawn, replete, he reenters the shelter of the woods he left thirty hours before. He crosses familiar scents and reorients quickly. Soon he is curled up, prepared in typical raccoon fashion to avoid facing the day, deep in the hollow chamber of the old loblolly pine.

THE TURTLE EGGS hatch in late summer, over a span of about twenty-four hours. Upon liberation from each eggshell, a baby turtle wiggles from side to side, pushing downward with his outsized, oarlike flippers. Each baby shimmies in the nest, moving upward until free of the leathery shell tatters and sodden sand.

In a chamber beneath a fragile canopy of barely moist sand, the babies congregate en masse. They wait until no more of their companions squirm up from below. Then they wait longer, until it is dark. Perhaps they know it is dark (without yet ''knowing'' what light is) because the air temperature above them drops to some critical level. We do not know; somehow, they do. Their nest erupts. Flailing their paddle-flippers, a hundred baby turtles burst from the cooling surface. They head downslope away from the dark dune grass toward the moonlit ocean. We have known that baby loggerhead sea turtles scull over sand, away from dark toward light, since the experiments of one Dr. Hooker in 1911. Their innate, newborn's tropism for the light tugs them toward the sea and the horizon, which, even without the moon, would be lighter than land.

On this night no raccoons patrol this strip of beach. Most birds have tucked in already. Only one, a night heron, announces herself by her vernacular name *qwok*. She is flying high over the marsh west of the road when she detects her prey. She rudders east, descending. She can spot the tiny turtles' ungainly motion a mile away. She knows what they are, exquisite morsels from the sand. The ghost crabs know, too.

Frantic, the baby turtles now face a gauntlet. Most will be a tough fight for any ghost crab. On tiptoe, zooming, a ghost crab angles into the stream of turtles, watching for one that pauses, and pounces. Snared by a foreflipper, a baby turtle must either quickly fling the crab or succumb to it—wheeling in ever-slower circles, propelled by its one free flipper. Ghost crabs are very patient. Some of these infants lose flipper tips, some lose eyes, some lose their lives before their first taste of salt water.

The night heron, *qwok,* planes into the stream, snatches babies right and left and flings them. Her sharp bill cleaves necks and flippers, shears off plastrons—bottom shells. Some of the bill-tossed babies right themselves and take off again, downslope, toward the reflecting water. They do not try to avoid the heron or the crabs; they only flounder, some wounded now, toward the moonlight on the sea.

Lightning illuminates the murderous, grisly scene; the late summer sky empties rain. Great warm dollops of water spatter onto the beach and its confusion of turtles, crabs, and heron. The crabs despise this sudden freshwater bath; most scuttle for the salty safety of their burrows. The heron rips off another flipper and scoops a bundle of turtle innards. Hunched, she seems briefly to contemplate the carnage. Then she gathers the wind under her wings and lifts south, toward the shelter of the woods, looking for a stately pine tree.

The sharp pellets of rain energize the remaining baby turtles, their ranks ravaged and depleted. The lifting tide brings a wave upslope toward them. First one, then three, then a dozen feel the water beneath flailing flippers. Grace-less thuddings on sand suddenly become smooth and grace-ful arcs in water. The swash hisses back, stranding a few babies again, but the next wave takes them, or seems to.

They disappear going forward, downslope, while the wave rushes the opposite direction, upslope.

The great, gray North Atlantic is no match for the hatchlings. Each measures less than three inches long and weighs less than one ounce—the size of a small sand dollar. The ocean can hurl 4,000 metric tons of water mixed with 670 metric tons of sand, along a 200-meter front, due west, up the beach, straight into the snouts of the baby turtles.

The turtles do not even know it is a contest. Their oar-flippers bite water. They swim—no, fly—directly into the sea, straight through her tumbling wave, due east. For the baby turtles, sixty days on land, almost all of those days inside an eggshell, have been a fleeting moment in life, a foreign time. That time was an anomalous oddment, an evolutionary holdover from the Paleozoic, more than 300 million years ago. They might swim a thousand miles before dinner. They are home.

CONSIDER, NOW, the components of this elemental tale. Remove any one ingredient—sea, wind, rain, raccoon, turtle, snake, heron—and you have a fundamentally altered story. Some players in the drama might have fared better, some worse. No matter: the same players will be back tomorrow in a different act. And the sand. The sand will always be there, the ultimate life-giving prop, pliable and plentiful and porous to all who traverse it.

That other prop, the sea, is changeable, too. By the time the turtle eggs hatch, the molecules of water and salt in the wave that brought the mother turtle have long since reached Cape Florida. Some have dispersed to the Gulf of Mexico; some, caught in the Gulf Stream, have gone to Scotland. The sand in that wave has dispersed, too. Some still lies in the interdune swale a mile north of Cape Hatteras light.

Some has found Georgia. The next time the mother turtle approaches the beach, heavy with eggs, she will again await the perfect wave, though it will not be the same wave. It will be in many ways not even the same sea. And how does she know which wave is perfect—which will carry her farthest up the beach again? How does she even know that there will be soft dunes in which she can lay her clutch? More questions, it would seem, for the poets.

Similarly, one can only intuit that the southwest wind that lifted those flakes of calcareous sand—flakes of a Cape Cod clamshell—and so successfully hid the turtle's nest, is not altogether the same wind that pushed Lieutenant Maynard's vessel toward Blackbeard in 1718, or lifted General Billy Mitchell, pilot extraordinaire, from the Hatteras sands in 1923. The same oxygen that refreshed Orville and Wilbur Wright in 1903 was perhaps manufactured on the coast of Brazil by a tree that had not even begun to grow when, in 1585, John White mapped the inlet, open from sea to sound, just north of the great maritime forest now called Buxton Woods. It was this inlet that time and sand had filled to a wading creek for a young raccoon famished for a water snake. It was that slip of water where, in fall, the night heron flew from a Buxton Woods loblolly pine to hunt baby water snakes in the marsh grass.

It requires a prodigious imagination to perceive in just which ways this sea is not that sea, this wind is not that wind, and this creek is in fact that once-great inlet—and will one day become an inlet (but not quite the same inlet) again. It may take scientists years to figure out where that water, that sand, those baby turtles go; and in the exercise they posit another set of equally tantalizing questions.

If tiny flakes of calcareous sand cannot unlock all the secrets, assume a vastly different perspective: from an Apollo

9 capsule 120 miles above the dunes. That is what astronaut Rusty Schweickart did at 10:00 A.M. on March 12, 1969, when he snapped a hand-held Hasselblad camera and captured the Outer Banks from Nags Head to Cape Lookout. Space lore has it that the Banks were the only landmark unobscured by clouds that day.

The now-famous photograph clearly shows the islands of Hatteras, Ocracoke, and Portsmouth and the Pamlico Sound that bathes them from the west. It shows the fans of sand where sound water hurries through the inlets out to sea. It also exposes a thin dune line that parallels the Outer Banks several miles inland. Known as the Suffolk Scarp, it is the geological remnant of an earlier coast—a sure sign that the Outer Banks will not stay put, as sea-level rise, relentless winds, and climate changes roll the islands westward.

Few can duplicate the astronaut's perspective. But anyone can stand on the beach, gaze out to sea, taste the salt, feel the wind's caress, watch the birds overhead, glimpse the ghost crab, and know that they are all connected, part of one continuum. It is a connection that has lured seafarers, scientists, tinkerers, pirates, explorers, and pilots to the Outer Banks since, in 1524, Giovanni da Verrazano sent twenty-five men ashore "in the latitude of 34 degrees," now believed to be an area between Cape Lookout and Cape Hatteras.

Looking for water, they found Indians instead. Verrazano thought the Outer Banks were an isthmus separating the ocean from the coast of India, China, and Cathay. Based on navigational knowledge of the day, it was a plausible supposition, one that some explorers clung to for another 150 years. Regarding expectations of the exotic and the unexpected on the Outer Banks, as with the Banks themselves, there is no end in sight.

Chapter Two

Land

When we first had sight of this Countrey, some
thought the first lande we sawe, to be the continent:
but after wee entred into the Haven, wee sawe before
us another mightie long Sea: for there lieth along
the coast a tracte of Islands, two hundred miles in
length, adjoyning to the Ocean sea, and between
the Islands, two or three entrances.
Captain Arthur Barlowe, 1584

Stretching along the North Carolina coast for more
than 175 miles, from the Virginia line to below Cape
Lookout, is a string of low, narrow, sandy islands
known as the Outer Banks. They are separated from
the mainland by broad, shallow sounds, sometimes as much as thirty miles in breadth, and are
breached periodically by narrow inlets which
are forever opening and closing.
David Stick, 1958

A most significant feature of barrier islands
is their pronounced tendency to move, particularly
by recession. This movement relates to their origin,
present circumstances, and the
forces that act on them.
Paul J. Godfrey, 1970

I N THE WINTER OF 1959, a young graduate student at Louisiana State University packed his Volkswagen and drove from Baton Rouge to Manteo, North Carolina, gateway to the Outer Banks. Though excited about his new assignment, Robert Dolan did not know that he was about to spearhead a revolution in the study of coastal geology. Nor did he dream that—teamed later with a similarly precocious young botanist, Paul Godfrey—he would challenge and eventually tame forces that can be as mighty as any glacier: the U.S. Army Corps of Engineers and the U.S. Department of the Interior.

In 1959 it was not immediately obvious that Bob Dolan would even become a scientist, much less a distinguished geologist. Until the late 1950s, geology had been a staid and static discipline devoted primarily to examining rocks, sediments, and strata. That did not seem a likely pursuit for a self-described Southern California beach bum, a dedicated surfer who slept under piers to avoid the twenty-five-mile drive to his inland home.

But any determined surfer becomes a keen observer of his surroundings. He watches the relationship between tides, waves, currents, and winds. He develops an intuitive feel for the way nature works. Dolan was nothing if not observant. It was during a subsequent stint in the navy, for example, that he observed the key difference between the men up on the bridge and the men sweating down in the engine room: those on top had college degrees.

Following the navy, Dolan enrolled at Oregon State University and concentrated in earth sciences. He also continued to visit the coast—this time on field trips—and worked as

18

a lumberjack in Oregon's vast forests. His experience in lumbering demonstrated, as he says now, that "man was an extremely powerful agent of landscape change." In the course of a month, men with their chainsaws and heavy equipment could gash a wooded mountain into a bald, eroding heap of dirt and rock. It was an impression that would stay with him for the rest of his career.

After Dolan earned a master's degree at Oregon State, where he honed his fascination with coastal geology, his major professor encouraged him to pursue his doctorate at Louisiana State University, a national leader in "soft rock" geology. It was good advice. It led budding geologist and erstwhile surfer Bob Dolan by circuitous route to the Outer Banks. It also thrust him into the thick of a sea change in geology. The dusty science was itself being rocked by intellectual earthquakes, the aftershocks of which are still felt today. The earth's surface, it turned out, is not a stable crust. It is made up of some fifteen granite plates, each of which drifts slowly on a mantle of partially molten basalt rock. As they tilt and move, the plates' rubbings can trigger violent volcanos and earthquakes.

The first lecture Dolan attended as a graduate student concerned plate tectonics. He and his colleagues learned quickly that their revered textbooks did not explain this flood of new data. Here was evidence of plates moving, of a planet Protean, not permanent, in shape. It was like boarding an airplane you assumed to be still on the runway, only to find it airborne at 30,000 feet. To survive, you had to hang on tight; you also had to question the wisdom of those below who had assured you the plane would be on the ground. These young geologists discounted much of what they read in their textbooks. They also discarded the conventional teachings of professors they would not have dared challenge just a few years before. If they were to advance in their chosen profession, they would have to chart their own course. They would have to defy conventional notions and reshape the science of geology—literally from the ground up.

Young Turk Dolan was well-suited to such adventure. When his major professor suggested he undertake a research project near Cape Hatteras, North Carolina, Dolan had to consult a map. When he saw that he would be surrounded by the ocean he loved, he enthusiastically accepted. His major professor, a geologist of international reputation, wanted to know why that thin strip of sand known as the Outer Banks did not wash away. He assumed the Banks were held in place by a series of Pleistocene coral reefs underneath. That would make the barrier island system at least 10,000 years old, probably much older. He wanted Dolan to go there and find out just how far down in the sand those reefs were.

Armed with a drilling rig, Dolan traversed the length of the Outer Banks and the adjacent mainland for a full year. He drilled holes in the Dismal Swamp, on Roanoke Island, and all the way south to Ocracoke. In all, he drilled 140 holes, some of them up to 100 feet in depth. In not one of them did he find evidence of a coral reef. In every case, though, he found that the sand went down for about 30 feet and that it was Holocene, not Pleistocene, in character, meaning that none of it is any older than 5,000 years— modern, or "immature" in geological talk. Beneath that first layer he simply found more sand, a finer variety belonging to an earlier Pleistocene terrace on which the Outer Banks sit. "The Outer Banks," Dolan says now, "are nothing but a little prism, a ribbon of sand, and it's only about thirty feet thick."

This might not seem a momentous discovery, but in

retrospect it was. It meant that the Outer Banks were of very recent origin and, given what was beginning to be known of sea level rise in the early 1960s, that the Outer Banks were not fixed islands but dynamic geological systems capable of moving, as winds and currents and violent storms dictate. For young Bob Dolan, it was a revelation and an opportunity. He was sure of his conclusions because he knew the subject firsthand, better than anyone else: he had lived every inch of it. "I've always felt that there wasn't a dune out there that wasn't my best friend," he says. When distinguished scientists came to the Outer Banks that year, they consulted him, not the senior researcher for whom he ostensibly worked. The usually rigid academic hierarchy was turned on its head; the junior graduate student was lecturing the doyens in his field because he knew more than they did, and what he knew was right.

It was a confidence that Robert Dolan never lost. Now an eminent professor of environmental sciences at the University of Virginia, Dolan is a senior scientist on a faculty of twenty-six, one of the best in its fast-growing field. Of that group, though, only six faculty members are considered generalists; the rest are specialists focusing on discrete pieces of the geological puzzle. Dolan, ever ebullient and confident, vividly remembers his first days in the field and curses the advent of the computer, which he calls "the worst damn thing man ever invented."

Instead of engaging in field work, today's graduate students can sit behind a computer screen and create "models" of geological systems, including the Outer Banks. "Now, you don't have to worry about drilling a hole," Dolan says. "You can simulate the damn thing." Dolan still counsels his graduate students that "there is absolutely no substitute for hands-on field experience." He also worries that the science of coastal geology has become too reductionist—that it has moved away from the broad overview that only the generalist can provide. With both indictments, that ultimate generalist and island-hopper, Charles Darwin, would surely concur.

How were the Outer Banks formed in the first place, and why? What makes them such perennial targets of debate in scientific and political circles? What can they tell us about the increasingly critical study of sea-level rise and global warming? To what extent are they a potential laboratory for studying fundamental questions of man's survival on this planet? The search for answers to these and related questions, to the extent that they can be answered, must begin with a good map of the globe.

Surely no one deeply interested in the earth or life upon it can avoid the fascination of maps. Maps draw the eye and tease the intellect. Similarly, no serious peruser of maps could be less than astonished by the Outer Banks, riding so far out to sea. Yet many thousands, perhaps millions, of people—beginning with the aboriginal Amerindians and continuing to the tourist hordes each summer—have actually visited the Outer Banks and have not been so astounded. Each year larger crowds adore the shellfish and game fish. They visit the Wright Brothers National Monument and read about Blackbeard the pirate. They note the commercial signs that commemorate Billy Mitchell. They are sometimes more bored than agog at the vastness, the sheer length of it all—and most never get near the lower half! But few ever contemplate the larger picture, the implications that only maps and basic geology can suggest. That picture has not been painted yet. What follows is a set of preliminary sketches that only further research can fill in.

At first glance, you will notice that the Atlantic coast of

North America is strikingly distinct from any other coastline on earth. It has elbows. Jutting rectangularly or acutely into the sea, separated by grand, scalloping indentations called bights, are America's bony elbows—the Capes of Cod, Hatteras, Lookout, Fear, Romain, and Canaveral. You will also note that the broader the expanse of lowland that lies inland of these elbows, the closer they are together. The closer they are to terrain made from glaciers, the bigger they are.

Our coast's uniqueness can be verified by further inspection of your globe or map. You will find peninsulas, of course, and sand spits and barrier islands on and along plenty of other coasts. But the discriminating viewer will quickly perceive that nowhere else on earth does this regular pattern of elbow capes occur. This is only a fairly recent discovery, and one whose importance has only recently been affirmed.

Indeed, until the arrival of Robert Dolan in the early 1960s, and the subsequent investigations by a motley collection of botanists, zoologists, and an occasional journalist several years later, scant attention was paid to this striking coastal phenomenon, even by the best geologists. It is almost embarrassing to think of the eminent scholars who devoted lifetimes to examining gravel beds, boulder heaps, and odd rocks, but who skirted the great elbow capes with all the apparent terror of the most skilled sea captain in a hurricane.

Begin now with the Gulf of Mexico. It is flanked by barrier islands and spits from Florida to Vera Cruz. They are large and well-developed in both directions from the mouths of the Appalachicola in Florida, and the Rio Grande, where Texas and Tamaulipas meet. But they do not ride far out to sea. They are outer, not Outer. Complex and as yet unexplained physiographic patterns of arcs and angles appear in the Gulf, too, especially from Mobile Bay to the Mississippi. But the Gulf can boast no great elbow capes.

The Pacific Coast of North America is steep. There is little continental shelf on which sand and sediments can massively accumulate. The geology is tectonic: buckling, folded ranges and sheer, plummeting faults even beneath the sea. Still, Northern California has a wonderful elbow cape, Point Reyes. Farther south, at San Diego and down in Mexico, are short runs of barrier spit, but they are of little consequence. On this coast the prevailing winds, westerlies, tend to push the sediments and sand onshore, not out to sea. This feature generally prevents major barrier systems from forming along the west coasts of all temperate climate lands. In these climes, Northern Hemisphere or Southern, the prevailing winds are always westerlies. The same conditions affect the coasts of Europe, West Africa, western South America, and Western Australia, too.

The Northern Hemisphere coast most closely comparable to our Atlantic coast is that of China. China has oak trees and alligators, *Spartina* marshes and garter snakes, greenbriar and ratsnakes. In fact, the south and east of China uncannily resemble the south and east of North America biologically. But the geology is utterly different. China's continental shelf is broad, but still steep and folded. It is a set of troughs and mountains culminating in a massive range of mountainous offshore islands: Taiwan to Japan, north to the Kamchatka Peninsula.

Billions of tons of sand and sediments do pour into the Japan, East China, and Yellow seas, but only small spits and barriers form. The currents are so disoriented and abbreviated by the flanking islands and the Korean Peninsula that lengthy formations cannot develop. Most of the sand and sediment simply fans out across sea bottoms. Also, North

America's elbow cape systems owe their existence, indirectly, to the mighty Gulf Stream. The Stream admittedly goes out to sea far off all but Cape Canaveral; but it creates counter-currents, called *gyres*, that curl back into the bights.

A Gulf Stream analog can be found in the Pacific: *Kuro Shio*, the Japan Current. But it is much weaker because so much surface water slips through the East Indies into the Indian Ocean. In the Atlantic the water escapes through the West Indies, all right, but when it hits the shores of the old Spanish Main, it all gets deflected north. And the *Kuro Shio* courses mostly outside the island rim. It might make elbow capes along Japan, for example, if Japan were inside a shallow continental shelf where sediments could accumulate. But it is poised on the shelf's outermost edge.

The warm gyres off the Gulf Stream, by contrast, touch the southerly shores of Cape Hatteras. They contribute to the formation of a distinct ecological line that runs just south of Hatteras Island's unique maritime forest, Buxton Woods. Above the line can be found species of plants and animals more typical of northern climes; below it, warmed by the Gulf Stream's waters and prevailing southwesterly winds, species of austral and subtropical bent begin to thrive.

Take a short boat ride due east of Cape Hatteras. Within an hour's time you will cross the boundary of the Gulf Stream, a vast underwater river within the sea. You will sense an unforgettable difference: the bluish green water, the fresh sea air, a greater abundance of sea creatures (or so Hatteras fishermen hope). Technically the Gulf Stream begins at Cape Hatteras and literally drops off the continental shelf as it meanders farther out to sea. Land's end marks the beginning of what William H. MacLeish, in his recent book about the Gulf Stream, calls "the great-grandfather of waters. . . ." Early explorers to America discovered that they could ride the Stream and its gyres to their advantage: farther out to sea if their destination was Europe or points north; hugging the coast to follow the southbound gyres to Florida and the Caribbean.

Next, examine the east coasts of India and Africa. Along the coastline of India proper, the Indian Ocean is creating elongated spit barriers that are rapidly building. This effect is a result of the massive erosion destroying the Indian subcontinent. It is only in the last century that egregious overpopulation has generated the land destruction that has built these barriers. They are still small by American Atlantic standards and are scarcely studied at all.

East Africa is little studied with respect to coastal geology. No notable spits, barriers, or elbow capes are evident. This no doubt reflects the narrowness of the continental shelf, along with the relative absence of large or numerous rivers feeding sediments to the coast. The climate is generally rather dry; there is no breadth of coastal plain.

Similar climatic and land features define eastern Australia. But this island continent does have a broad, gently sloping continental shelf. This shelf has developed not so much barrier spits and islands—although a few can be found—but the Great Barrier Reef of coral. Nothing approaches it in the Atlantic. That leaves only one other exotic coast to examine: the east coast of South America.

There it is: the longest running barrier spit system on earth. Spanning 700 kilometers—twice the length of North Carolina's Outer Banks—and equally thrust into the sea, they are the Outer Banks of Brazil. But these Banks are arcuate; they lack elbow capes. They also have only one inlet, centrally located. This inlet is stable. The Outer Banks of Brazil do not fragment into barrier islands.

≈

WHEN ROBERT DOLAN proved conclusively that the Outer Banks were nothing more than wind-swept strips of sand thirty feet deep, he unlocked a key to barrier island systems that had eluded scientists for generations. While it is true that many renowned geologists ignored these complex systems, not all of them did. Indeed, it could be argued that the early European explorers who set foot on these dunes were the first to give them rudimentary scientific examination. They had to, in order to survive, and to discover new routes to the exotic East.

Several of them mapped the islands and their inlets and wrote of what they saw. Notable among these were Philip Amadas and Arthur Barlowe, dispatched by Sir Walter Raleigh to the New World in the spring of 1584. Described by Outer Banks historian David Stick as "America's first English-language publicist," Barlowe wrote effusively of these shores, from "The highest and reddest Cedars of the world" to the Indians, "very handsome, and goodly people. . . ." But the most meticulous work was left to artist John White, whose several trips to the Outer Banks—including his shepherding of the fateful "Lost Colony"—have left a rich store of maps, drawings, and descriptions of England's first attempts to settle in the New World.

But it was not until nearly 300 years later that scientists began to ponder the origin of these peculiar offshore protrusions. It was a Frenchman, Élie de Beaumont, who first studied the role of wave action in building barrier islands. De Beaumont's studies, published in 1845, led to a later flurry of activity by, among others, American G. K. Gilbert, who is credited with first using the term "barrier." But the dean of early shoreline experts was a Columbia University scholar, D. W. Johnson, whose 1919 book on the subject became a classic in its field.

In retrospect, Johnson was more right than wrong in his theories about how barrier islands form. He coined the term "offshore bars" to describe them. He also developed a concept called "profile of equilibrium" to describe how the slope of a beach and adjoining shallow ocean bottom adjusts naturally to absorb the energy that waves expend on it—from the gentle lappings of a summer ocean to the huge winter surges that periodically wash over barrier islands.

Johnson deserves credit, too, for detecting the importance of inlet formation, longshore currents (which move along-side the beaches), and direct wave action in shaping barrier islands. Johnson was also ahead of his time in concluding that typical barrier islands are not fixed and stable, but capable of being pushed toward shore by wave action. In this conclusion he stood against no less an authority than the famed Harvard naturalist Louis Agassiz, who believed (along with Robert Dolan's major professor years later) that the offshore bars of the southeastern United States rested on coral reefs below. In Johnson's day, it was even seriously theorized that the barrier bars were not created by the ocean, but by the flow of subglacial streams.

Where Johnson veered off course was in assuming that barrier islands are created by a *retreating*, not a rising, sea. Under his theory of the "shoreline of emergence," as the sea retreats its forward waves build up barrier bars. Those bars are pushed toward shore and eventually merge into the existing beach. In his text, Johnson imperiously rejects the theorizing of a small number of geologists (and at least one botanist) who said that only a rising sea could explain the behavior of a barrier island. As a result of Johnson's domi-nance in his field, the forward progress of coastal geology was frozen for many decades.

What these early studies lacked, until the perfection of

carbon-dating techniques in the 1960s, was confirmation that barrier islands were very recent in origin and that they were changing. Carbon dating merely corroborated what Dolan had found with his drilling rig. It also helped uncover what Dolan calls the missing "linchpin" of barrier island study: barrier island sands are not only capable of movement by wind and wave, they are also affected by a gradually but inexorably rising sea. Without an inherent ability to move as sea level rises, the Outer Banks would not be a complex ecological system of beaches, dunes, hammocks, and marshes. They would be one of the world's most spectacular underwater sand parks.

To UNDERSTAND WHY these barriers, spits, and elbows are among the most dynamic geological systems on earth, and how sea-level change affects them, we must develop a time-lapse picture—one of continuous, variable-speed change. A hundred thousand years ago the ocean level stood twenty meters, or seventy feet, above its present level. The current position of the Outer Banks was accordingly underwater. But Outer Banks existed even then. Wave action and sedimentation produced the Penholoway Terrace, well inland from the sea's edge today. Flanking this terrace, just offshore, were the Outer Banks of the Sangamon Interglacial Period.

About 80,000 years ago the sea began to drop. Its retreat occurred because more and more ice began to collect at the earth's poles. That ice quite simply consumed available water: took it literally out of circulation from the sea. Sea level fell in stages; each set of barrier banks was in turn stranded, high and dry, on the widening coastal plain. When sea level finally stopped falling—about 50,000 years ago at the beginning of the great Würm Ice Age or Glacial Maximum, a submarine continental shelf no longer existed. No Outer Banks and no elbow capes occurred along the Atlantic coast of North America because there was no gentle slope on which they could form. Land extended some thirty kilometers—nearly twenty miles—east of Cape Hatteras and twice that far east of Cape Lookout. There, sea bottom plunged as precipitously as it does off Japan today.

At the Würm Glacial Maximum, the edge of the continental ice sheet came to the present position of Long Island, New York. Wooly rhinoceroses and mammoths tramped the plains where shrimp trawlers ply the sounds today. Giant ground sloths, bigger than grizzly bears, munched trees on the broad coastal savannas. It was a very different world.

About 12,000 years ago more ice began to melt at the edges of the glacial ice caps than accumulated on them. Sea level began to rise. By 10,000 years ago the Würm ice was in full-scale retreat. Inundative sea level rise advanced far more rapidly than had been its rate of fall during the post-Sangamon period. The world changed quickly. Concordant with coastal inundation and rapid climatic warming came waves of hungry people. The giant animals of the Ice Age— wooly rhinos, mammoths, wild horses, giant sloths, even saber-toothed tigers—could not meet the combined onslaught. They soon became forever extinct.

With sea level rise, the Outer Banks formed. Melt water fed great rivers that carried rock flour, sand, and gravel across the narrowing coastal plain and nourished the fledgling Banks. First spits, then islands, then elbow capes developed. Plants and animals took up residence on these narrow new lands. New ecological communities developed, too— perhaps closely paralleling communities that existed during the Sangamon Interglacial 100,000 years before.

It is tempting to speculate that those ancient Sangamon

barrier island plants and animals may have survived on the sandbars stranded on Würm coastal plain. Perhaps genetic constitutions and ecological relationships persevered to be recaptured by the freshly formed Outer Banks, evolving into what we know and see today. Small species may well have survived the almost incredible changes of the late Pleistocene and recent millenia. Some of them may outlive us yet.

About 3,500 years ago—an instant by the clock of geological time—sea level rose very rapidly to a point about two meters (a fathom), below its present position. The climate then cooled and the rate of rise slowed down. The sea did not halt its rise. Despite the name "little ice age" given to this cooling trend, nothing resembling a genuine glacial period occurred. But this pause in the pace of inundation, combined with the flood of outwashed sand and sediments, permitted the great barrier systems to build.

Today you can see the earlier dune systems fragmented along the mainland edge. Part of this ridge is faintly outlined in the astronauts' Apollo 9 photograph. Harkers Island corresponds to old Shackleford Bank. A previous Cape Lookout made an elbow at Harkers' eastern end. Just south of Stumpy Point in Pamlico Sound lay the primordial elbow of an old Cape Hatteras. Sand ridges formed rapidly to seaward and, despite the still-rising sea, broke the surface to form parallel series of dunes. For a time, sand accumulation was so rapid that the Banks swelled longer and broader despite the sea's advance. But eventually the rate of outwash slowed; new sand gain fell behind inundation, and the situation stabilized.

Perhaps "stabilized" is the wrong word. The Outer Banks are a geological system; stability implies a structure. Sand continued to move continuously and often rapidly. Inlets opened and closed, inlet (really outlet) deltas formed. The low ground overwashed and dunes marched inland. Through all this change, however, for at least a millenium, the positions of the elbow capes and barriers remained roughly the same. It was a restless, tenuous stability.

IN THE THOUSAND years prior to A.D. 1800, sea level rose only about twenty centimeters, or eight inches. But with the advent of the Industrial Revolution and an exponential increase in human populations, the climate began to warm up again. The release into the atmosphere of so-called "greenhouse" gases from man's consumption of fuels—coal, wood, gas, oil—is adding layers to the thin, atmospheric blanket that traps the sun's heat and warms the earth.

The principal greenhouse gas, carbon dioxide, is not poisonous; plants require and use it in huge quantities. It enables them to grow, and in turn to feed us and (directly or indirectly) all other animals. So why is it that the world's plants do not literally bloom on all our artificially produced carbon dioxide and mitigate the greenhouse effect? Because, historically, man has relentlessly destroyed the earth's vegetation, especially the big and the old. From such forests are lawns, fields, and parking lots made.

A forest of mighty trees consumes countless tons of carbon dioxide and converts it to plant tissues. By contrast, the wispy blades of a wheat field can consume a scant few pounds. Thus the carbon dioxide accumulates and begins to work the way glass works in a greenhouse: it allows the heat to come through, but it retards mixing, cooling, and the escape of heat back out to space. As human populations increase, we cut down ever more forests to plant crops, or house more people, or park more cars. We burn more fuel, too; and we are figuratively cutting off the limb that supports us.

A major consequence of this pattern of global warming is that the polar ice caps are melting more quickly. Scientists are not sure how fast this is occurring, or how the earth's climate will respond to the greenhouse effect. But that sea level will rise faster is clear. Since the year 1800 sea level has come up at least thirty-five centimeters, or fourteen inches. Most of that rise has been documented in the last 100 years, since 1890. Sea level is rising about three inches, or eight centimeters, a decade now, and it will inevitably accelerate.

All manner of apologists for business-as-usual erect all sorts of arguments against the obvious: humans are causing rapid climatic warming and rapid sea-level rise. The greenhouse effect is indisputable.

As sea level rises, the positions of the elbow capes and barrier spits and islands will change. The continental coastal plain will be submerged. The capes will shift southward. The barriers will move inland. No matter what man does, there will be Outer Banks. They may move all the way to the positions they held in the days of the Penholoway Terrace of the Sangamon Interglacial. They might move farther inland than that. Ancient sea terraces can be found 250 to 270 feet above present sea level, cut by Miocene seas some 20 million years ago. All the world's ice might melt. There would still be Outer Banks. They will move, and with them will go their natural inhabitants: dune grass, salt marsh, birds, rats, and snakes.

For us, and for many other species, the repercussions will be cataclysmic. But nature will triumph no matter what we do. Instead of the customary picture of the survivor of nuclear holocaust on a desert island, the realities of the late twentieth century demand a new cartoon: a survivalist camper on the Outer Banks, ready to pick up his or her tent and move to the highest dune as the waves wash over everything, flooding our beach-front homes and our hubris—the naive and arrogant thought that we could stand up to the sea's inevitable summons.

Chapter Three

Water

It is melancholy fact, Sir, that on this dangerous
coast, beset in all directions by shifting shoals and
changeable currents . . . the Engineer . . . is never
safe while his only dependence is on the
perfection of his charts and quadrants!

William Tatham,
surveyor, 1807

Most people who live on a barrier island watch the
weather and seasons more closely than those who
live inland. . . . The one time of the year when
Ocracokers are especially attentive to the weather is
during hurricane season.

Alton Ballance, 1989

B Y 1962, ROBERT DOLAN'S Outer Banks research had evolved from the relative crudity of drilling holes to a more sophisticated project. The experiment's purpose was to measure the impact of storms on a specific section of the beach and nearshore. Armed with a grant from the U.S. Navy, which took an interest in his work, Dolan spent seven months and $50,000 setting up his experiment. He rented a 650-foot pier in Nags Head to install his equipment. Finally, in mid-February 1962, his forty-day study began.

It did not take nature long to oblige him. Twenty-one days into the investigation, a ferocious storm stalked the East Coast. It was one of those late-winter northeasters that periodically strike the Banks, bringing gale-force winds and powerful storm surges. But this storm, in the 100-year category, nearly went off the charts: it packed seventy-five-mile-an-hour winds—hurricane force—and it arrived with the perigean high tide, one of the year's highest.

On the morning of March 7, after a night of howling winds, Dolan peered out the window of his oceanfront cottage, set back about 200 feet from the beach. Water was racing across the property. A vacant cottage in front of him, closer to the beach, was bobbing toward him in the raging surf. It was time to beat a prudent and hasty retreat. Dolan took refuge in a two-story motel up the beach. Water was already flooding the first floor when he arrived. From his vantage point on the second floor, he saw cottages being dashed to pieces by waves and debris in the surf. Dunes were being washed over. Roofs were being ripped off.

Hours later, after high tide had receded, Dolan returned to his research site. Much of the beach road had been layered with up to five feet of sand. Huge overwash deposits had been carried 300 to 400 feet inland. Dolan's fishing pier had been chopped up, fodder for the storm's fury. His thirty-foot camera tower had completely vanished. Two-inch-thick steel pipes that he had driven deep into the beach as markers were twisted and broken like pipe cleaners. The sand around them had eroded up to five feet. The barrier dunes behind the markers had been sliced back more than 120 feet. Later in the day, six miles down the beach, Dolan found the pier pilings to which he had attached his wave and tide gauge. Miraculously, its graph paper could still be read. Electric power had gone out just after the gauge had recorded a monster wave measuring eighteen feet in height and lasting seven seconds.

It was a turning point, both for science and the Outer Banks. The Ash Wednesday Storm, as it was soon called, spread destruction up and down the East Coast. The storm caused $300 million in damage; hundreds were injured and many died. On the Outer Banks, lagoons were filled with overwashed sand. Dunes were badly eroded. An inlet new to us, but right where John White found Chacandepeco in 1585, had been carved open north of Buxton on Hatteras Island. Roads and driveways were buried.

The Ash Wednesday Storm washed over Bob Dolan with the force of revelation. It demonstrated how quickly and ruthlessly a storm can rearrange barrier islands, as if they were pieces in a child's block set. But after the storm subsided, Dolan also noticed that the basic rhythms of wave, wind, and current returned to their normal pace. Despite its devastating effect on humans, the storm was part of a larger geological pattern or system. Far from being destructive, it had a constructive effect. Sand had not been washed away, but had been pushed over the dunes by overwash. The total "budget" of sand in the system had increased, not decreased. The beaches, if anything, became wider and more gently sloped. Dolan became convinced that overwash was the dominant pattern in barrier island geology, and that man's attempts to build up artificial dunes to block the ocean's advance were futile and counterproductive.

One serendipitous effect of the Ash Wednesday Storm was that it increased the flow of scientists and research dollars to the study of coastal geology. Among those interested in learning more about beach dynamics were the National Park Service and the U.S. Army Corps of Engineers. The Park Service worked out a research arrangement with Dolan, who continued his studies of the Outer Banks from his new base at the University of Virginia.

The Corps was called in to help with beach restoration after the Ash Wednesday Storm, and to devise ways to minimize property damage during the next storm. The first order of business, for example, was to close the new inlet north of Buxton. That consumed $4 million and took a year, and it was only the beginning of a series of expensive Corps proposals to make the Banks safe for human endeavor.

THE ASH WEDNESDAY Storm was one of the most awesome of this century, but it was hardly the only major tempest to batter the Banks. Violent weather is as integral a part of life on the islands as fishing, crabbing, and surfing. The Banks and their plant and animal denizens are well-adapted to foul weather: they simply submit to it. For human inhabitants, however, submission is genetically more difficult. The tendency of humans to settle in one place and to defy the elements invites conflict, not harmony, with nature. Those who do make the Banks their home must be resourceful, and they must be risk-takers. The general rule of thumb is that if your tolerance of perfectly freaky weather is low, the Outer Banks are not the place to settle down.

The warm summer of 1990, for example, brought numerous sightings of waterspouts—water-borne tornados. Although waterspouts are a fairly common occurrence in the waters off the Outer Banks, several of these spouts had close brushes with people. One whirled ashore less than a mile down Ocracoke's main swimming beach. It kicked up sand and startled sunbathers before dissipating. Another summer spout lifted a camper from his tent on Hatteras Island and dropped him, nude, into a ditch 100 feet away. Though shaken and scratched, the camper was not badly hurt. News reports did not say whether the man's state of undress merely preceded, or was caused by, his rude intruder.

Ocracoke ferryboat captain Rudy Austin well remembers when a waterspout slammed into his ferry early one morning. It was foggy; Austin and his crew had no warning. He heard a humming sound, like a tornado. Fortunately, there were no cars or passengers on the ferry. The cyclone rocked the vessel and spun trash out of the garbage cans. Had the cans not been lashed to the boat, they would have been carried aloft, too.

While dangerous, waterspouts are isolated disturbances. Hurricanes, on the other hand, mean business. They swagger. Regarding hurricanes, Rudy Austin considers himself lucky. Despite the affinity of violent weather for the Outer Banks, it has been a long time since the area absorbed a direct hit from a hurricane. In 1985, the center of Hurricane Gloria, headed for landfall on the Outer Banks, abruptly turned north at the last minute and averted disaster. Even so, the flood water at Rudy Austin's place in Ocracoke village licked the top of the third cinder block, just under his front door. Twice he has jacked up his house to add another cinder block. After Gloria, and after reviewing the devastation of Hurricane Hugo on the South Carolina coast in September 1989, Austin wishes he had gone higher. It is at times like these that Outer Bankers sense the vulnerability of their existence. If a Force 4 or 5 hurricane were to slam into the Outer Banks, there would be no escape, no higher ground. The islands would be submerged. The results would be devastating, at least for human inhabitants.

But Rudy Austin is right: Outer Bankers have been exceedingly lucky. No hurricanes have made landfall on the Outer Banks since the 1930s and '40s. Yet the odds of a hurricane hitting the islands remain good. Researchers report that since 1900, the Gulf and Atlantic coast barrier islands have been crossed by more than 100 hurricanes. The fury of Hugo, a Force 4 hurricane that made landfall at Charleston, can now be added to that list of undesirables. In the summer of 1991, Hurricane Bob passed just east of Hatteras but caused little damage.

Hurricanes and northeasters have played a critical role in both the geological and human development of the Outer Banks. In the 1580s, violent storms repeatedly steered the English away from a permanent settlement on these shores. In June 1586, a vicious hurricane roared through Roanoke Island. As historian David Stick reconstructs the tale, this event proved a pivotal moment in the history of New World colonization. Sir Francis Drake, military hero and world explorer, had just arrived to bring supplies and reinforcements to a small garrison of English settlers left on the island the previous fall. He offered the would-be colonists a choice: either he would leave them ships and supplies for another month of exploration, after which they could decide whether to return to England; or they could sail back with him immediately.

The hurricane quickly narrowed the garrison's options. According to the journals of Ralph Lane, leader of the colonists, Drake "in that storme sustained more perill of wracke then in all his former most honourable actions against the Spaniards." An observer aboard Drake's vessel reported that the violent storm brought "thunder. . . and raigne with hailstones as Bigge as hennes egges." Also sighted were "greate Spowtes at the seas as thoughe heaven & [Earth] woulde have mett." The vessel Drake had assigned to Lane had been lost, with several men aboard. Supplies were now severely limited. Several smaller boats had been smashed. Again, Drake offered Lane a choice: stay with a single vessel and fewer supplies, or return to England forthwith. Lane's men had seen enough. They would leave.

Had the hurricane not intervened, it is likely that Lane's band would have kept a toehold on the island. Had they done so, they would soon have soon received more supplies and reinforcements from ships dispatched by Sir Walter Raleigh. Though delayed, those ships were moving up the coast as Lane and Drake departed; they arrived soon after the weary colonists left. Stick speculates that "Had the relief vessel reached its destination more expeditiously with relief for Lane, the continuity of settlement on Roanoke Island would, in all probability, have been maintained." In other words, the Outer Banks, rather than the Chesapeake Bay to the north, might have become the site of the first permanent English colony in the New World.

Or would it? Exposure to foul weather was not the only factor that militated against Roanoke Island's selection as a permanent colony. The shoals and sandbars in the inlet were notoriously treacherous. Large vessels could not enter the sound. Men and supplies had to be offloaded at sea and transported to the island in smaller boats. Even in the best weather, these boats frequently ran aground and capsized. With its deep harbor and protected shoreline, the Chesapeake would prove more hospitable to human habitation.

But the emergence of Jamestown as the first permanent colony was preceded by one further, futile attempt to tame the Outer Banks: John White's Lost Colony. Roanoke Island was not White's first choice for the colony. But circumstances beyond his control forced him to settle there in July 1587, rather than move north to Chesapeake as planned. Again, inclement weather played a role in the colony's brief and mysterious existence—and disappearance.

A summer northeaster blew through in August, forcing Simon Fernando, the ill-tempered pilot accompanying White, to abandon his anchorage and move out to sea. The storm's nastiness made Fernando impatient to leave and hastened White's difficult decision to abandon his daughter and new granddaughter, Virginia Dare, and return to England to plead for reinforcements. But it was not until the summer of 1590, three years later, that White was able to return.

When he finally returned to the Outer Banks, by now desperate to learn the fate of his family and friends, White was once again frustrated in his search. As he and his fellow shipmates headed for Roanoke Island in their small boats on August 17, another northeaster whistled in. The vessel in which White rode half-filled with water, but the craft made it safely to shore. A second boat was not so fortunate. Menacing waves in the inlet capsized and overturned it. Seven of the boat's eleven crew drowned. As Stick wrote of the incident, "It was the forerunner of many other disasters that were to occur in succeeding centuries along that isolated stretch of coast destined to become known as the 'Graveyard of the Atlantic.'"

White and the remaining crew finally reached Roanoke Island later that day, but it had become too dark to search for the colonists. The next morning White found none of his group—only their settlement abandoned and overgrown with weeds, and the word CROATOAN carved on a tree or post at the entrance. That was the signal agreed to between White and the colonists to indicate that they had moved farther south, to Croatoan on Hatteras Island.

White and his pilot, Captain Abraham Cooke, agreed to sail the short distance to locate the missing settlers. But more bad weather that night tossed their ship violently. By morning the anchor was riveted in the sand. The anchor cable soon broke; the ship drifted dangerously close to the beach, where it would have been dashed to pieces. A second anchor cable snapped. Only a third and last anchor stood between the crew and shipwreck. This one held; miraculously, the ship slipped into a deeper channel, a slough, and Cooke was able to manuever her seaward. Still, White never reached Croatoan. Dwindling supplies and an understandably skittish crew demanded a change in plans. After enduring more violent weather at sea, Cooke and White returned to England. The 116 men, women, and children left behind three years before would not be found this trip—not by John White, and ultimately not by anyone.

Thus did the Lost Colony become "lost": because efforts to *find* them never materialized. When White did not return in good time, it is not unreasonable to suppose that the colonists moved south to Hatteras. They had discussed this option with White before he left, and they used the agreed-upon sign. White assumed they had sought more favorable conditions at Croatoan.

What happened to the colonists afterward is a matter of endlessly fascinating but fruitless speculation. Did they die of disease? Did the Indians kill them? Or, as seems more probable, did some of them eventually die of various causes while others survived and mixed with the Indians? These questions will probably never be answered. What can be verified is that the inhospitable environment of the Outer Banks, from the islands' treacherous shoals to their spectacular storms, stymied Raleigh's first attempts to colonize the New World.

As an historical footnote, that changeable environment has also frustrated more recent archaeological diggings to locate the Lost Colony settlement. While remnants of Ralph Lane's fort have been found, no trace of the separate settlement has been discovered. But Robert Dolan and two colleagues believe they have a plausible explanation: the probable settlement site is more than a quarter of a mile out in Roanoke Sound, buried under water. The northern tip of Roanoke Island is especially vulnerable to wave erosion, by virtue of its exposure both to northeasters and to surges from strong low-pressure centers—a combination found nowhere else along the Outer Banks. Dolan estimates that between 1851 and 1980, 960 feet of shoreline on the northern end of the island were eroded. Assuming similar rates of erosion in preceding years, the island's northwest point would have receded 2,000 feet on average since the Lost Colony was founded. The proposition does have one flaw: it is hard to explain why the settlers would have located their living area forward of, rather than behind, the fortress intended to protect them from enemy attack. Nonetheless, there is melancholy satisfaction in speculating that the physical remains of this ill-starred settlement, its inhabitants so profoundly affected by vicissitudes of wave and wind, lie buried in an underwater grave.

The Banks' reputation as a great leveler of ships and dreams has only magnified in the centuries since European explorers first attempted to settle there. David Stick has counted more than 500 shipwrecks off barrier islands from Cape Henry, Virginia, to Cape Lookout. Some ships were sunk during wartime battles, but the majority have been lost during sudden storms. Most feared of all were the Diamond Shoals off Cape Hatteras—properly named "the Graveyard of the Atlantic." Positioned far out to sea, these shoals mark the convergence of two great ocean currents, the Labrador from the north and the Gulf Stream from the south.

Among the famous vessels lost in this vicinity was the Union ironclad *Monitor*, which went down in a storm while being towed off Cape Hatteras in December 1862. Four officers and twelve men were lost. In 1974, Duke University scientists found the ship about sixteen miles south of Cape Hatteras. Subsequent diving operations have yielded some of its contents.

The Outer Bankers' responses to shipwrecks have often been heroic. From lifesaving stations along the shore, Bankers habitually rowed out in high surf to rescue passengers and crew, and then welcomed survivors into their homes.

But legend also holds that some Bankers tricked passing ships by tying lanterns to the bobbing heads of ponies and running them along the beach. The ships' captains were supposed to mistake the lanterns for the lights of other ships safe at harbor—and sail toward them. When they ran aground near shore, the vessels were then boarded and looted. The tale conveniently clears away the mystery surrounding the naming of Nags Head. And it is plausible to suppose that scavengers profited from the contents of ships' cargoes washed ashore. But some observers, including David

Stick, are more inclined to believe accounts that the high dunes of Nags Head reminded early sailors of a familiar promontory called Nag's Head on the coast of England.

STORMS–HURRICANES and northeasters—are the most spectacular architects of barrier islands, because they move huge quantities of sand so quickly and visibly. But they are not the only agents of change at work. In fact, sand moves in five major ways. The degree to which one or the other of these five patterns of sand transport performs its handiwork determines the physiography of the Outer Banks, generating spectacular differences from one island to the next.

Longshore current transports sand inexorably from north to south down the Banks. Longshore current is generated indirectly by the Gulf Stream, which flows northeast across the Atlantic. Small countercurrents peel off from the Stream and carve into the northern ends of the bights, just below each elbow cape. A roughly analogous phenomenon takes place with the Gulf Stream in the North Atlantic, but on a much larger scale. A great deal of the Stream's water curls back westward across the North Atlantic, hits North America, and—made ice-cold by its long journey—is deflected south. The ocean's relentless need to remain level produces longshore current. It moves sand southward, elongating spits and barrier islands at their southern ends and attempting to close inlets.

Outflow from the rivers crossing the mainland constantly pushes sand (and silt or mud) into the sounds. Outflow currents concentrate at what are called "inlets." As David Stick pointed out long ago, that appellation is wrong: the gaps in the Banks are outlets. Because the net flow of water is out the "inlets," huge quantities of sand and sediments are deposited. Outflow, like longshore current, tries to close off inlets by forming inlet deltas and shoals.

Prevailing winds, *southwesterlies*, blow tons of sand northeastward. The blowing sand accumulates around vegetation, or any object, and builds up. Even where no objects exist, the sand blows into dunes. Until the dunes fetch up on some immobile object, they tend to move northeastward.

Storm winds can come from any direction. The really big storms—the hurricanes and northeasters that plagued the early settlers—are cyclonic. Centered on a low-pressure cell, they draw wind and cloud into themselves in a counterclockwise spiral. When the low-pressure cell lies over water, the warm water nourishes and maintains it, supplying the heat and moisture to keep the pressure low. The capacity of cyclonic winds to move sand is enormous. The northeasterlies are the most effective winds. They come into the low cell from top right—northeast. If the cell is centered over Pamlico Sound, the intensity of the incoming wind and its great oceanic fetch can in minutes reverse the effects of years of prevailing southwesterlies.

Overwash is the aquatic counterpart of storm wind. It occurs only during the great storms. While overwash can occur from the west, this effect moves relatively little sand and is really a component of the outflow pattern. The really muscular sand movers come in from the sea. Given a combination of easterly gales, low pressure, storm-generated giant waves (called swells), and high tide, the ocean carries right over the Banks—and carries millions of tons of sand with it.

The first three sand movers—longshore current, outflow, and prevailing southwesterlies—are regular and relatively gentle. They do their quiet work in spring and summer. The last two—storm winds and overwash—work in violent and sudden concert. They effect spectacular changes the tourist sees

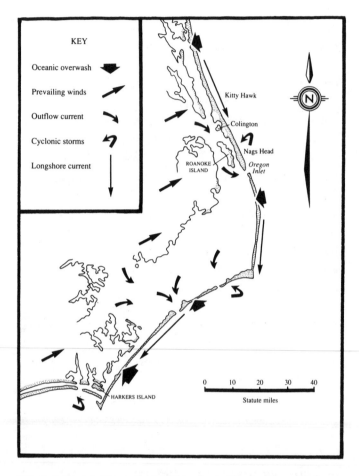

Wind and water, Outer Banks. The stippled land areas are Holocene sediments believed to be less than 6,000 years old. Note the "islands" of old Pleistocene land at Kitty Hawk, Colington, Nags Head Woods, and Roanoke Island. Remnants of former Holocene barriers can be seen at Harkers Island and in fragments along the mainland. Arrows indicate major features of water and wind. Overwash can occur anywhere, but carries across to the sounds only where the Banks are narrow. Prevailing southwesterly winds are the norm except in winter. Outflow current results from rivers discharging into the sounds. Cyclonic storms include hurricanes and northeasters. Longshore current is generally southward, but short northward currents occur at Oregon Inlet and Cape Henry in Virginia.

only upon returning the next year, after fall and winter have done their leveling.

The relationship of each barrier island to these forces dictates its individual size, shape, movement, and recent geological history. A gull's-eye tour of the islands from north to south shows how. From Virginia to Pea Island the Banks run northwest to southeast—athwart, normal to, or at right angles to both the prevailing southwesterlies and the northeasters of great storms. These winds work in opposition. The storm winds blow huge amounts of beach sand across the land toward the sound. The prevailing winds push it back toward the beach. The result is the development of the highest dunes on the Outer Banks: the towering hills and wondrously steep slopes of Jockey's Ridge and environs, which face into the wind. They do not occur elsewhere.

On the north-south run from Pea Island to Cape Hatteras, sand movements achieve a kind of steadfast normality. All five sand movements would proceed quite naturally but for *Homo sapiens*, who maintains a paved highway and some clusters of buildings here. Storms would open new inlets, deltas would form in them, overwash would fill the soundsides, and sea-level rise would move the Banks westward.

David Stick historically tabulates three inlets in this stretch. New Inlet, or Chickinacommock, periodically opened and closed from about 1730 to the 1930s. When it was open, it made Pea Island an island. Loggerhead Inlet, just south (and just north of Rodanthe), probably opened about 1846, when Oregon and Hatteras Inlets opened, and was closed by 1880. Chacandepeco, just north of the large east-west land mass at Cape Hatteras, perpetually tries to open again today. It was open when John White depicted the Banks in 1585, but it was closed by 1657. Outflow accumulation at this narrow neck of the Banks has had a bearing on the development of the next piece of Banks—the Buxton highlands and Buxton Woods.

Often overlooked by tourists and increasingly desecrated by residential development, Buxton Woods in many ways constitutes the most distinctive region of the whole Outer Banks. North to south, this land mass is about four miles wide; it spreads almost due west from the Atlantic just north of Cape Point for about eight miles. In all, the Buxton highlands add up to more than thirty square miles.

The geological development of this bit of the Banks has been unique. Longshore current carries sand down off Cape Point, Hatteras Island, and out to sea. The Portsmouth Bight gyre curling inland off the Gulf Stream picks up some of this sand and moves it westward, then southwest again, along the Banks: another longshore current. Outflow piled up a grand delta to the north. Prevailing winds picked up beach sand and blew it northeast across the east-west barrier. The sand, replenished from the shoals and tip of the Cape, marched inland (northward) in rather low, but very broad, parallel dunes. Storm winds from the northeast moved sand from the spit above Chacandepeco Inlet, but their overland (or over shoal and marsh) fetch mitigated their effect on the dunes marching north from the bight.

Vegetation rapidly held and perpetuated this system as much as 3,000 years ago. The prodigious amounts of sand constantly being drawn south and west into the bight, only to be blown north and east onto the land, assure that—left to nature—the Buxton highlands would grow southwestward apace of any erosion on the east and north resulting from sea-level rise.

≈

FROM CAPE POINT the Outer Banks veer southwest all the way to Cape Lookout. This region is the Intercapes Zone. It is unique because here longshore current moves sand southwestward in direct opposition to prevailing wind. As elsewhere, sand in the water is moved down the Banks, but at low tide a lot of it dries out and gets blown right back up—northeast. This constant conflict between the two most incessant forces of sand movement means dunes do not build to grand heights, and spits accrete more slowly. This in turn means that inlets tend to remain open for relatively long times.

However, the longer an inlet stays open, the more effectively it dooms itself. Huge inlet deltas form in long-lived inlets, so that one day—when closure is finally achieved—hugely widened land areas result. Inlet deltas undergo ecological succession from sand shoals to eel grass flats to marshes. Once vegetation breaks the water's surface, storm-blown sand begins to accumulate very quickly. A patch of marshland, buried by blowing sand, becomes a dune-grass prairie. Waterbush (*Baccharis halmifolia*) quickly takes root, and in its turn knocks down even more blowing sand. Soon yaupon (*Ilex vomitoria*) and wax myrtle (*Myrica cerifera*) can grow and, accumulating more sand, begin a small freshwater reservoir. Yaupon, especially, thrives on sand burial. Bury a yaupon trunk and each branch sprouts roots and becomes a new, small tree.

Ultimately, inlet deltas succeed to hammocks dominated by live oak (*Quercus virginiana*) and redcedar (*Juniperus virginiana*). That the biggest, most long-lived inlets make the biggest inlet deltas, which in turn make the biggest hammocks, is a set of geological facts with the most profound implications for man and nature, as we shall soon see.

The last link in the Outer Banks chain is Shackleford Bank.

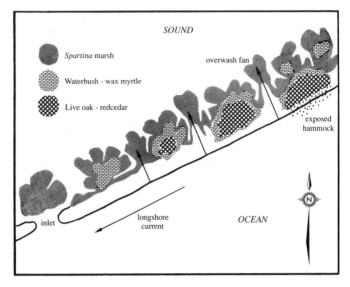

A diagrammatic view of physiographic processes in the Intercapes Zone that accompany rising sea level. Beginning lower left (southwest), *Spartina* marsh forms on an inlet delta. Longshore current extends the point northeast of the inlet, while outflow eats away the point southwest. As windblown sand accumulates, old inlet delta marshes fill, allowing waterbush and wax myrtle to take root; hammocks then form with live oak and redcedar. Overwash fans build soundside marsh and maintain island width. Rising sea level cuts back the seaward face of the island, eventually exposing stumps of old hammock trees and often engulfing the living ones.

It is analogous to the Buxton highlands, running east-west and, in a state of nature at least, densely forested. Today Barden Inlet cuts Shackleford from Cape Lookout. But sand moves constantly westward across the inlet's southern end and attempts to fill it.

Most of the differences between Shackleford and Buxton land areas derive from the deforestation of Shackleford. This artificial deforestation was perpetuated by use of this bank for livestock, left to run wild or become feral. Because no one has lived on Shackleford for most of the last century, the livestock were uncontrolled until the 1950s. Ecological succession will eventually make Shackleford Bank like the Buxton highlands again. It will take centuries, however.

IN THE LATE 1960s, another young scientist had made his way to the Outer Banks, by a route slightly less circuitous than Robert Dolan's. Paul Godfrey, a botanist and a native of the midwestern plains, earned his Ph.D. at Duke University in 1968. Like Dolan, he was drawn to the Banks, but to a much less-traveled strand of the ribbon: the new Cape Lookout National Seashore, authorized in 1966. The park's first superintendent wanted a complete ecological study of these islands—Portsmouth, Core, and Shackleford Bank, along with Cape Lookout—and Godfrey was the right man for the job. With his wife, Melinda, who was already working there, Paul Godfrey spent the next two years checking historical records, taking core samples, and performing experiments with various grasses to see which ones best withstood the Banks' harsh environment.

Godfrey quickly concluded that the Banks are receding toward the coastline, that storms play a critical role in their regression, and that the chief agent of change is overwash, which pushes sand from the beach face to the sound side of the island. As a botanist, Godfrey recognized that the dominant plants in the islands' ecosystem were quite well-adapted to overwash and salt spray. Indeed, he found that in the apparent destruction from storms and overwash are contained the seeds of the islands' survival. Overwash fans on the sound side of the islands become nurseries for new grasses, which in turn trap sand and create new marshlands.

In 1970, Godfrey published a thin but important document in the history of Outer Banks research: *Oceanic Overwash and Its Ecological Implications on the Outer Banks of North Carolina*. In thirty-six pages of type, charts, and photographs, Godfrey helped precipitate a debate that still rages in some quarters today: whether man's attempts to stabilize the dunes, beaches, and inlets are essentially futile and injurious to the natural forces that would otherwise govern them.

Godfrey did not mince words: "By attempting to hold everything in one place," he wrote, "man is actually creating a much more unstable situation that will lead to greater problems of erosion in the future. Erosion is a man-conceived evil that only man worries about, especially when it threatens his structures. . . . In the long run, he will make the situation worse because such interruptions will put the islands in greater jeopardy of destruction as long as the sea continues to rise. Man is the barrier island's greatest enemy, not the sea."

At the time, Godfrey presumably could not have imagined that those simple words, and others like them, would reverberate through the scientific community and ultimately reverse longstanding policies of the United States government. "We must protect these islands from the works of man, not the sea," Godfrey declared. But who was he, a mere postdoctoral researcher, to challenge the authority of the Army Corps of Engineers and the National Park Service?

The Corps, after all, had been a familiar presence along the Outer Banks since Congress appropriated $20,000 in 1828 to purchase a dredging machine to render Ocracoke Inlet navigable for vessels. By 1834, Army engineers had worn out two dredges and were working on a third; yet their labors had yielded few visible results. By the time channels were dredged from sea to sound, their mouths would have already shoaled up again. Over the course of the winter, whole channels would simply disappear. Finally, in 1837, Congress appropriated additional funds to build a jetty to protect the Ocracoke channel. But just before the jetty was completed, a violent gale destroyed it. Thus, between 1826 and 1837, Congress had spent nearly $133,750 to stabilize the inlet, and had nothing at all to show for its largess. Work on the project was abandoned until 1890, when renewed efforts to dig a channel once again failed miserably.

Despite these setbacks, the Corps of Engineers persisted in its efforts to help coastal residents and commercial interests protect their property from storm damage and promote naval traffic through the Banks' treacherous inlets and sounds. When the National Park Service (NPS) began to manage much of the Outer Banks as a national seashore in 1937, the Corps of Engineers was called in for advice. Workers for the Works Progress Administration–Civilian Conservation Corps (WPA-CCC) built fences to catch sand. In the 1950s, the Park Service engaged in extensive dune stabilization by creating a high dune ridge along the length of Hatteras and Ocracoke Islands. Its primary purpose was to protect vulnerable stretches of the newly completed Highway 12, which traverses the length of the Outer Banks from Ocracoke to Corolla. The decades of the 1950s, 1960s, and early 1970s, meanwhile, were notable for the frequency and severity of northeasters and hurricanes, from Hazel (1954) to Ginger (1971). Between 1958 and 1972, in fact, Congress appropriated more than $10 million for beach erosion control, including more than $1 million to repair damage done by the Ash Wednesday Storm.

Shortly after Paul Godfrey had arrived on Core Banks, the Park Service proposed another expensive dune stabilization and beach nourishment project. But the government's case had been weakened. Powerful members of Congress questioned the advisability of throwing good money after bad. Nonetheless, $4.3 million was appropriated in 1971. And that sum was only the beginning. That same year, the Corps estimated that stabilization of the entire Cape Hatteras National Seashore shoreline would cost $50 million initially and $2.5 million in annual maintenance. The idea was to create a big dune, 20 feet high and 500 feet wide, all along the Outer Banks.

To Dolan and Godfrey, these plans were ludicrous. Both began to speak out boldly against the stabilization effort, within National Park Service circles and at scientific conferences. Dolan recalls that when NPS officials asked him what he thought of the plan, he told them "it was the stupidest thing I'd ever heard of."

In 1971, before Godfrey assumed a new post at the University of Massachusetts, he and Bob Dolan met for the first time. It must have been a highly charged rendezvous. Both men realized that they had arrived at the same conclusions independently—working on different sections of the Banks and peering through different scientific lenses. They agreed that barrier islands should be allowed to function as natural systems, without human interference, and that the Park Service should be persuaded to embrace that philosophy.

Thus began one of the most potent collaborations in the history of coastal research. Together, Dolan and Godfrey set out to stop the largest long-term coastal management project in history. The odds were formidable. They were arrayed against a battery of government experts on the other side who had a management plan for every barrier island from Fire Island to Georgia. But, as Dolan notes, he and Godfrey were armed with the most powerful weapon ever devised:

the truth. Quite simply, they were right. The evidence was all on their side, and they had studied it meticulously. "If you're right," Dolan argues, "it doesn't matter how much ammunition the other side has." These two young assistant professors made speeches, attended meetings, and wrote scientific papers. Their ideas were adopted and spread by others. A 1972 article in *Science News* endorsing their theories went out to a half-million readers.

In the end, no other scientist disputed their findings. The National Park Service had to admit that past stabilization policies had not worked. A major study of dune stabilization published in 1973 concluded, "There is no reason to assume that the strategies for management we have inherited are the best or the only ones for dealing with shoreline problems. These systems change rapidly. Most of our past decisions for park development were based upon the implicit assumptions that we can extend our stable 'continental' world down to the edge of the sea. We now know this is impossible, both in terms of environmental degradation and simple economics. . . ."

Dolan and Godfrey had won. Not only did the Park Service accept their arguments, it acted upon them. In 1973, Cape Hatteras Superintendent Robert Barbee confirmed the Park Service's abandonment of any attempts to stabilize the barrier islands. "We can't continue to proceed as we have," he said. "We're not God, we can't hold back the sea forever." The announcement made national headlines and the national news.

That conclusion is still disputed, but it has held, at least on Park Service land. Elsewhere—in the Corps of Engineers' proposal to build large jetties to stabilize Oregon Inlet, for example—efforts to subdue nature along the coast have not died out. But Dolan and Godfrey achieved what few scientists are ever able to achieve: a marriage of science and public policy that reversed a decades-old way of doing business. Their campaign against barrier island management had reverberations that extended far beyond the Outer Banks. If dune stabilization was harmful to the Outer Banks, the Park Service calculated, it must also be harmful to other barrier islands up and down the East Coast. Best to leave nature in charge of them all.

In the intervening years, Dolan and Godfrey have gone their separate ways. Dolan is constructing what he hopes will be a grand design or model of coastal dynamics. Godfrey is no longer engaged in field studies of barrier island ecology. The battle against coastal development has been taken up by geologists such as Orrin Pilkey, Jr., of Duke University and Stan Riggs of East Carolina University. Refinements of the old theories of overwash and beach erosion have been proposed and accepted.

Riggs, a veteran of North Carolina coastal studies, is working on a new explanation of the Outer Banks' formation that is far less dependent on overwash, wind, or longshore currents. University of Maryland professor Stephen Leatherman, a former colleague of Godfrey's at the University of Massachusetts, has been able to develop a synthetic overview that combines new information on overwash with the older views on longshore accretion and inlet delta formation.

Bits and pieces of theories going all the way back to de Beaumont and Johnson—the right mixture—have found their way into a consensus overview that Lazell, John A. Musick of the Virginia Institute of Marine Science, and their students first articulated to Godfrey in 1971. Hypotheses that seem diametrically opposed often turn out, with the cooling influence of time, to merge quite comfortably.

Early observers established the thesis that degradation of plant communities by humans and their livestock had accelerated storm damage and further reduction of natural ecosystems. Their cure was to rebuild and revegetate a dune rampart along the seaward flank of parts of the Outer Banks. Godfrey and Dolan developed the antithesis: overwash is natural, plant communities are adapted to it, and—therefore—artificial dunes are evil. As late as 1975 both schools seemed to be utterly beyond reconciliation; but it turns out both were quite correct in large measure. A synthesis emerges: human activities had indeed harmed natural communities, so some repair was advisable; and overwash is perfectly natural and helps to build natural communities.

The results of the synthesis can today easily be observed on Ocracoke. The island was nearly all deeded to the Park Service, and human and livestock degradation ceased by about 1965. A massive, continuous rampart "dune" of sand was pumped into place along the seaward flank, and planted in northern dune grass (*Ammophilia breviligulata*), which is naturally scarce south of Cape Point. This wholly artificial construction gave the shrubby and woody vegetation a respite from overwash for a few years, and excellent recovery resulted. Today, storms have breached the rampart and grand overwash fans extend northwestward across the dune and shrub zone, greatly building Ocracoke up in many places.

At the same time, the woodcutters and stock animals vacated Portsmouth Island, just southwest of Ocracoke. Here, however, no artificial rampart was built. By 1970 the difference in the two areas was dramatic: Ocracoke had far more shrub and woodlands; Portsmouth remained rather bleak.

By 1989 the distinction between the two had blurred

dramatically. Overwash was reclaiming some shrublands as prairie on Ocracoke, while the extensive bare sands of Portsmouth had in many areas succeeded through the prairie stage to shrublands. The high ground at Portsmouth by 1989 supported a fine growth of redcedar woodland in extensive patches, and live oaks were coming back in numbers where, in 1971, you would have had trouble finding one.

The developing picture is one in which the natural (technically "subclimax") condition of the Intercapes Zone supports extensive, if patchy, areas of hammock woodland interspersed with prairies developed on overwash fans. The old argument about whether the "natural" condition of the Banks was one of "continuous" forest is irrelevant now: forests were far more extensive before the arrival of axe and cow, but always breached here and there by prairies, marshes, and grassy dunes. Ocracoke reverted to an approximation of this condition more quickly than Portsmouth, but at great financial cost. Once the degrading activities of humans are curtailed, nature can handle the reconstruction, given time, as Portsmouth of today is proving.

But the development of a synthesis in no way detracts from the accomplishments of the authors of the antithesis. For sheer daring, the contributions of Dolan and Godfrey to our understanding of the Outer Banks cannot be surpassed. Their work celebrates the power of ideas to topple the twin peaks of government bureaucracy and scientific orthodoxy. It also illustrates the perpetual lure of the Outer Banks to inquisitive adventurers, from Lane and White to the exploits of Dolan and Godfrey.

Chapter Four
Blackbeard

Piracy, being a crime not against any particular state
but against all mankind, may be punished under
international law. . . . But, whilst the practice of
nations gives to every one the right to pursue and
exterminate pirates without any previous
declaration of war. . . , they may not be
killed without trial except in battle.
Phillip Gosse, 1957

Edward Teach died a violent death, but it was in the
heat of battle, as he would have wished, still
fighting as he fell with the insensate rage of
a mortally wounded lion.
Robert E. Lee, 1974

In real life . . . , most pirates probably led miserable
lives. They were often drunk, and quarrelsome.
They wasted food, liquor, and money. Many
pirates died of wounds or disease.
The World Book Encyclopedia, 1989

T HE MOVEMENT OF sand under the influences of wind and water has predicated the histories of men on and around the Banks in dramatic ways. One of the most colorful and famous incidents in Colonial history, the killing of the pirate Blackbeard, Edward Teach, in 1718, played as it did on a stage set by geological forces and shifting sand. The battlefield on which he died was the culmination of centuries of outflow, hammock formation, channel movement, and shoal development. Blackbeard capitalized on the results of these processes to great advantage. He failed to carry the day—and save his own head—because of his own personal failings. Nature afforded him every chance of success, even outnumbered two to one.

Blackbeard's nemesis, Lieutenant Robert Maynard, won the day with presence of mind and pluck. But Blackbeard drifted into complacency when he believed sand and shoal had given him victory. He simply celebrated too soon.

The configuration of land, channels, and submarine shoals benefited those, like Blackbeard, who took the time to learn them, and kept their knowledge constantly up to date. Because of the ways sand moves, things are not often the way they seem to the sailor navigating these waters. Away from land does not always mean deeper water. Cutting close to a strand may be the wisest course—at least at a given moment, for the sand will not lie still.

A long history of spit growth, inlet delta formation, hammock succession, and oceanic overwash constructed the stage on which the battle played. Longshore current, outflow, and a great cyclonic storm were the stage crew. Great supporting roles were filled by prevailing southwesterlies and

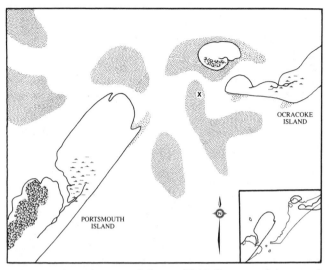

Ocracoke Inlet and surrounds in A.D. 1066. Portsmouth hammock is already well established, although the island here extends far northeast of its present tip. Ocracoke hammock is developing on a succeeding inlet delta. X marks the spot, awash on a shoal, of Blackbeard's future abode. Inset shows overlay of Ocracoke's present shape.

the turning tides. All these combined to make high drama of what might have been a simple massacre. But it was the human stars who directed the final outcome. On that fair November morning of 1718, success and failure, life and death, would be determined by the characters of the two men whose destinies were shaped by the unique geology of the Outer Banks: the actions of wind, water, and sand—and the greatest of these was sand.

Ocracoke Inlet is the longest-lived inlet known to man on the Outer Banks. It existed when John White painted the Banks in 1585 and, we believe, for hundreds of years before. It has shoaled and filled in the classic manner of an Intercapes inlet. Had shoal, channel, and land been in slightly different configuration—give or take a few years, for instance—the onset and likely outcome of the battle would have been wholly different. One of the great stories of the age of piracy would not have transpired as it did.

The high ground of Portsmouth Island has probably been high, and naturally covered with hammock vegetation, for a thousand years. It was no doubt once an inlet, and filled and succeeded—as do they all. But Portsmouth serves as the gate pillar of Ocracoke Inlet on the southwest. It has not in recorded history ever overwashed, and it resists erosion from rising sea level rather nicely. A thousand years ago, low ground very likely extended northeast of the Portsmouth hammock for about four kilometers, or more than two miles, into what is Ocracoke Inlet today. The inlet, then new, was similarly displaced to its position near the town of Ocracoke today.

At the time of the Norman Conquest of England, A.D. 1066, more than 600 years before Blackbeard was born, an aboriginal Ocracoke Point looked rather like its descendant of today. It was about six kilometers—more than three miles—northeast of its present position. The beach front of Ocracoke was about sixty meters seaward of its present edge.

At this time much of what would become the high ground, or hammock, of Ocracoke village, was a marsh island in the inlet delta, formed in the perpetual manner of sand moving under the influences of outflow and storm winds. The place where Blackbeard would, in about 650 years, build his temporary abode lay underwater on building shoal. In the accompanying map, in which it is shown how things might have been in 1066, the position of Blackbeard's abode is marked with an X.

During the next 500 years Portsmouth's low, northeast

spit eroded and submerged. Ocracoke point grew southwestward, largely by longshore current sand movement, and captured the shoal where Blackbeard would one day live. Capture of this shoal deflected the outflow channel northward long enough to scour out Cockle Creek, a narrow body of water that, in 1931, would become today's Silver Lake in the town of Ocracoke—ringed with small shops, piers, and motels. Although Silver Lake looks like an obvious and prominent feature of Ocracoke village now, it is largely a recent artifact. In Blackbeard's day it was a scarcely notable little estuary probably filled with marsh.

The Cockle Creek channel soon filled from outflow, perhaps with the help of overwash, and no doubt in concert with blowing sand. The channel was then deflected north again, to round the growing Ocracoke hammock. The hammock would effectively gate that channel to the present day.

To trace the inlet's history, we must turn again to early maps of the New World, which show a classic example of how an inlet evolves and matures. Because Ocracoke Inlet is the only inlet that has survived intact to the present day, we can more easily visualize what it must have looked like in Blackbeard's time. In 1585, artist John White depicted the Outer Banks in a lovely map, preserved today in the British Museum. At this time sea level had risen less than twenty centimeters (eight inches) from its position in 1066. But windblown sand had filled the sound side and built the inlet delta hammock lands. Land extended farther west of Blackbeard's abode than it does today. (White's depiction of Ocracoke bears a remarkable resemblance to the configuration of Hatteras Inlet in mid-twentieth century.) John White called the island and its inlet *Wococon*. That name stuck through 1666, when a map was prepared for Robert Horne and printed in London. It shows Wococon, and

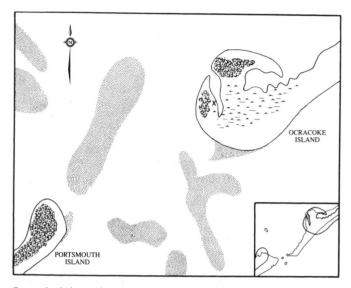

Ocracoke Inlet and surrounds, depicted as Wococon by John
White in 1585. **X** marks the future site of Blackbeard's abode.
Inset shows overlay of Ocracoke's present shape.

"C. Hope" where Cape Lookout is today, but offers no
useful detail of the inlet.

   In the map by John Speed, printed in London by William
Garrett in 1675, the sole innovation is the transformation of
the name of *Okok*. In 1682 a map printed for Joel Gascoyne
and Robert Greene, in London, and dedicated to the Earl of
Craven, backslides the name to *Wosoton*. This map clearly
shows three small islets in the inlet, no doubt a frequent
phenomenon, often repeated today.

   John Thorton and Will Fisher produced the map of what
they called *Wossoton* published at Tower Hill, London, in
The Fourth Book of *The English Pilot*, in 1689. This clearly
highlights the bulge of Ocracoke hammock and the projec-

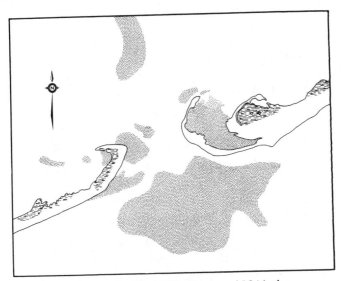

Hatteras Inlet. Just after World War II, around 1946, the physiography of Hatteras Inlet bore a striking resemblance to that of Ocracoke Inlet in 1585. The major differences are that Hatteras is not gated on each side by high, wooded land like Ocracoke and Portsmouth hammocks. And Hatteras forms shoals mostly on the ocean side, by longshore drift and storm surge, because it is so far from the Neuse River, a major source of outflow sediments. The island is quite ephemeral. This map amalgamates features of USGS topographic quadrangles "Hatteras" and "Green Island," 1946, and USGS chart 1110 for 1941 and (seventh edition) 1945.

tion of Ocracoke Point. It also shows Bluff Shoal approaching Ocracoke across Pamlico Sound from the mainland, as it always has.

John Thorton had already collaborated with Robert Morden and Phillip Lea in 1685 to produce a map that similarly shows bulge and point at Ocracoke. This map is highly stylized and depicts all ocean fronts as scalloped and indented, as they certainly never were. Nevertheless, the rendering of the shape of the end of Ocracoke has merit, and it is incorporated into our reconstruction of events at the time of a great storm.

A close copy of the 1685 map was made by Pierre Mortier and printed in Amsterdam in 1700. It is surely almost pure plagiarism. But Mortier may at least have troubled himself to ascertain if conditions at Ocracoke, the most important inlet anywhere in the Banks, were the same as in 1685. His map is here accepted as evidence that they were. Thus, sometime between this date and 1718, when Blackbeard was killed, it is likely that striking changes took place because of a storm.

Plausible evidence of such a devastating storm does exist. It might have been the great Carolina hurricane of 1713, which roared through on September 5 and 6. Eyewitnesses reported that the storm flooded the fortifications of Charleston, South Carolina (then Charles-town), and took the lives of seventy people there. Observers said the storm was even more violent farther north. A surveyor who passed through the Carolinas a few years later wrote of a sloop being driven three miles inland over marshes in the Cape Fear area, 145 miles south of Ocracoke. If not this storm, then certainly another of like force must have pounded the Outer Banks during this period. We will call it hereafter the Carolina Storm.

The storm altered the island's geography in dramatic ways. A vast overwash fan filled most of the marshland present on the sound side in John White's day. It made rather high ground of the swash between the beach and hammock. The runoff from this overwash scoured the Old Slough out into a prominent estuary. It remained as Ocracoke's harbor for the next century. This body of water, mostly marshland today, cut just south of Blackbeard's hammock, or Springer's Point. Its confluence with the main outflow channel round-ing Ocracoke hammock—and cutting off the western edge of high ground at Blackbeard's—made Teach's Hole, where Blackbeard would soon anchor.

In an abominable map made by Johannes Loots and printed by T. Jacobsz in Amsterdam in 1706, Ocracoke is called *Wocoton*. Loots could not have been more wrong: he confounded Cape Lookout with Cape Fear, thus abbreviating the Carolina coast by some hundred miles. He shows nothing of Core Banks and Portsmouth except a few little islets, the northernmost surely best interpreted as a shoal isle in Ocracoke Inlet. Ocracoke itself is shown short and fat, but its southwest end—the area of chief interest here—looks remarkably appealing. Because Ocracoke was so important— the only major, reliable inlet in those days—Loots may actually have known what it looked like.

Loots shows Ocracoke Point and the bulge of high ground. One may speculate that he truly meant to show the western, channel shore of Blackbeard's, or Springer's Point, somewhat cut away. Similarly, one may speculate that the breadth of the island Loots depicted northeast of the bulge had been filled by overwash. If so, this suggests the great storm may have occurred between 1701 and 1706. Yet Loots' failure to show Old Slough is equally plausible evidence that the storm had not yet occurred, giving credence to the 1713 Carolina hurricane hypothesis.

That great changes occurred is indisputable. All worthwhile subsequent maps show the area of water on John White's map well filled in, and the Old Slough as a prominent water body. The great overwash fan can be discerned on topographic maps of today; its age can be judged from its vegetation and ecological succession.

In our reconstruction of Ocracoke at the time of the Carolina Storm, the huge overwash fan and the beginnings

Historical depictions of Ocracoke Inlet and surrounds. (1) Land outlines reconstruct the inlet shape around the time of the Carolina Storm, 1701–1717: *bold line*, Johannes Loots map, 1706; *fine line*, Pierre Mortier map, 1700; *dotted line*, Thorton, Morden, and Lea map, 1685. (2) The bizarre configuration of John Lawson, 1709.

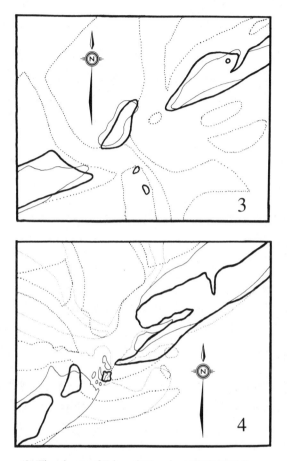

(3) The charts of Edward Moseley, 1733: *bold line and dotted lines*, land and shoal, respectively, from more detailed inset map; *fine line*, land shape, from main map. (4) The charts of Captain Wimble: *bold line and bold dots*, land and shoal, respectively, as published in 1738; *fine line and dots*, land and shoal, from 1733 manuscript probably based on observations and soundings around 1720.

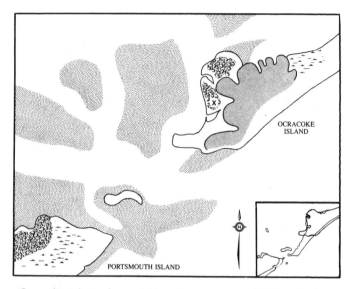

Ocracoke Inlet and surrounds at the great storm of the period 1701–1717. The massive overwash fan builds high ground for the future town site. **X** marks the soon-to-be site of Blackbeard's abode. Inset gives present Inlet shape.

of Old Slough are shown. The storm made the Slough much bigger and shifted Ocracoke Point southwest again. The storm also carried countless tons of sand into the inlet and produced massive shoals.

After the storm the main outflow channel was tightly restricted by the now vast shoal to the west. This channel cut the western face of Blackbeard's hammock to much the form it bears today. Teach's Hole, at the mouth of the Slough, was a fine, sheltered anchorage, and Edward Teach—Blackbeard—moved in.

≈

THE MOST CRITICALLY important historical reconstruction of Ocracoke Inlet, of course, is for November 21–22, 1718, when the Battle of Ocracoke began. Fortunately, we have an assortment of maps and charts for the relevant time, and enough recorded history to judge the skill of the mappers and charters.

John Lawson produced an artistically lovely map for *A New Voyage to Carolina*, published in London in 1709. Lawson shows *Ocacok* as a rather short, plump island with a large inland lagoon connected to the sound about midway on its northwest side. This lagoon extends up-island as an apparent disjunct pond just northeast of the creek's end. The whole configuration looks remarkably like a hybrid of modern Silver Lake, little Cockle Creek until 1931, and Old Slough, whose entrance, at the southwest end, was Teach's Hole.

Lawson's *Ocacok* displays a point not unlike the real Ocracoke Point, but truncated. Just off the point lies a large islet that looks remarkably like what other cartographers indicate as shoal. It begins well into the sound and extends well out to sea. A second islet lies off Portsmouth. The clear implication is that the inlet passage would be between the islets, rather than between the first, northeast big islet and Ocracoke Point.

That Lawson's lagoon was indeed a hybrid of Old Slough and Cockle Creek is a reasonable assumption. One can almost envision the salts of the day sitting down with tankards, a slab of pine, and some bits of charcoal, with which each sketched out his version of how the place looked—all at quite a geographic remove: no one lived at Ocracoke then.

The next cartographer on the scene is by far the most significant. James Wimble left Hastings, Sussex, England, in

1718 and sailed to the Bahamas. When he reached Ocracoke is not precisely known. But it was surely by 1721, because he was already then working on charts and maps. By 1723 he owned a square-mile section on the Carolina main in Scuppernong country near the present town of Columbia. According to his chronicler, William Cumming, Captain Wimble was to produce "the best coastal chart of the region until the end of the eighteenth century." Wimble's chart of 1733—probably researched, especially for Ocracoke, between 1719 and 1723—is authoritative for the time of the Battle of Ocracoke. Also convincing are some of the refinements Wimble made in his published chart of 1738, which is extraordinarily detailed and accurate for the whole region.

Why did Wimble's charts take so long to appear if much of his exploration occurred between 1719 and 1723? Because of Edward Moseley. By the time Wimble had achieved the prominence needed to earn any publisher's notice, Moseley was hard at work on the charts he was to publish in London in 1733: *A New and Correct Map of the Province of North Carolina*. Moseley did well with the Cape Fear and Cape Lookout regions, their inlets and their shoals. He seems to have had little knowledge of Ocracoke, however, and probably scarcely visited it. He provides two depictions, both on the same sheet. His ostensibly detailed inset is only about 17 percent bigger than Ocracoke Inlet on his main map. There are obvious discrepancies.

Moseley's *Ocacok* has a point, a bulge, and a creek or slough. The latter, however, is shown as about five miles northeast of the Point, closely conforming to the position of the Island Creek/Shad Hole system. Indeed, his inset shows a "well," probably a freshwater pond, in just the right position to be part of this system. Either Island Creek, now largely filled in and well succeeded to woods, or Shad Hole, rapidly disappearing today, would probably have been navigable to small boats in the early eighteenth century. Moseley calls all of the northwest side of Ocracoke, inside the big shoal, "Thatches Hole." This includes the area off what we take to be Island Creek. Thus Moseley's view was simply a hybrid or composite, inexactly recording the true situation in an untrue combination.

Fortunately, we have Captain Wimble. His 1733 sketch shows *Oakerccok* with point, bulge, Old Slough (just where they would be expected), an anchorage that must have been Captain Teach's, and the basic pattern of the shoals. By 1738 Wimble's *Okerccock* is even more refined and detailed. He located Moseley's odd creek and places it where Moseley did, and where it must truly have been if it is Island Creek or Shad Hole (both parts of the same system). The point, the bulge of the hammock land (now with buildings), and the anchorage in Teach's Hole off Old Slough are just about right.

Wimble shows shoaling off Ocracoke Point and what seems to be a small islet at the end of that shoal. Of course, the shoal may have broken water only at low tide. That scarcely matters: no one was going to sail over it, wet or dry. Wimble shows Royal Shoals and the shoals and channels northwest of Ocracoke's bulge. He depicts the Middle Ground just as Moseley did, right where Lieutenant Maynard spent a most uneasy night at anchor on November 21, 1718.

Wimble maps the great shoal extending south and east from Beacon Island, just as it typically forms and shifts since first drawn by John White in 1585. Incidentally, Beacon Island of Wimble and Moseley (and Lawson, for that matter, though unnamed) is not the same as the little shoal islet of marsh called "Beacon Island" today. Aboriginal Beacon

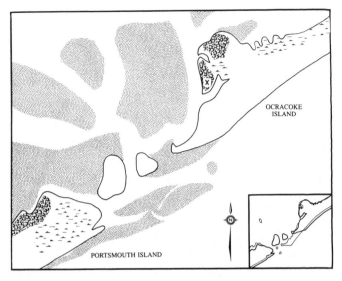

Ocracoke Inlet and surrounds on the day of November 21, 1718,
eve of the battle of Ocracoke. X marks the spot of Blackbeard's
abode. Inset compares 1718 Inlet shape with modern Inlet, below.

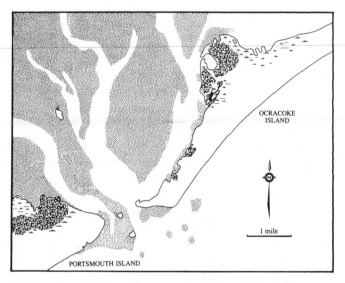

Modern Ocracoke Inlet and its surrounds. The canal development
and marsh destruction on the east side of Ocracoke hammock are
not shown, nor is the extent to which the town now dominates
the wooded hammock.

Island moves and changes, but is still more or less there today, in the inlet. A check of the sailing directions given by Wimble and Moseley confirms this point.

THUS DID CENTURIES of flow and surge, wind and storm, holding and gating hammocks, and shifting sands place shoals, channel, house site, and anchorages just where Lieutenant Maynard of the Royal Navy was soon to find them as he approached from the sea in the setting sun of that fair November day in 1718.

Maynard had to anchor off. He would have been a fool to navigate the inlet sailing upwind into a setting sun, or in the twilight. He did not know either the waters or the land, and he had come to engage in battle the consummate master of both. There was nothing Maynard could do but wait for the dawn. And Blackbeard must have known Maynard was there. From his hammock's high edge, near Springer's Point of today's charts, anyone could see the masts, if not the hulls, of Maynard's sloops across the sand flats and beyond the beach dunes. Moreover, Secretary Tobias Knight, in league with the pirates and Carolina's Governor Charles Eden, had written Blackbeard to expect trouble. Yet Blackbeard had no reason to be especially apprehensive just because two sloops had anchored off the inlet. The sloops might be merchantmen, harmless travelers, or Blackbeard's comrades in the informal Brethren of the Coast: fellow pirates. Had not Blackbeard just hosted at Ocracoke some of his notorious colleagues—men like Israel Hands, Charles Vane, Robert Deal, and John Rackham—in what is billed as "the largest pirate festival ever held on the mainland of North America"? (Thirty miles off the main, to be exact.) This party broke up in late October. Dubbed by Robert E. Lee in his *Blackbeard the Pirate* the "summit meeting of the

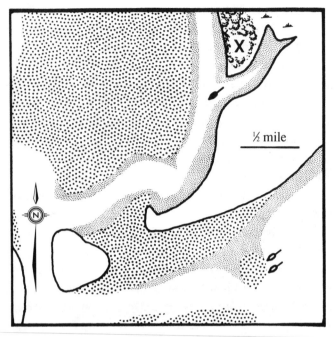

Sundown, November 21, 1718. Blackbeard's sloop *Adventure* (black) lay at anchor in Teach's Hole, the mouth of Old Slough. Lieutenant Maynard anchored his sloop and *Ranger* (both white) on the Middle Ground, outside the Inlet. Waves break at low tide on the shoals shown in heavy stipple. Fine stipple indicates sloop-grounding shallows.

elite of coastal piracy," it performed Blackbeard the disservice of being so prolonged and popular that everyone now knew exactly where he was.

Indeed, Blackbeard's reputation preceded him. He was both horrible and horrifying, no doubt of that. He killed his share of men and loved a larger share of women. For dramatic effect in battle, he braided his flowing beard with ribbons and stuck long, flaming matches under his hat. His large frame towered over the smaller men of his day. He carried a cutlass in his belt and three pairs of loaded and cocked pistols across his chest.

Nonetheless, the line between piracy and heroism was thin in Blackbeard's day. The great explorers now revered by schoolchildren were not averse to boarding a ship and stealing its cargo. Drake was one of the worst. He even participated in slave trading. But when these men looted ships, they were called "privateers" and enjoyed the blessing of the royal sovereign. That is because they were authorized and encouraged to plunder ships belonging to enemy nations.

Like most pirates, Blackbeard's career had begun legitimately enough. He was said to have distinguished himself in battle during the War of Spanish Succession, which ended in 1713. But he gravitated inevitably toward piracy. After settling in Nassau and acquiring his own vessel, he scored a number of impressive victories in the Caribbean before moving on, in January 1718, to terrorize the Carolina coast.

Blackbeard shifted his operations to Ocracoke in October of that year. As we have seen, with its shallow inlet and protected anchorage, Ocracoke afforded Blackbeard an ideal vantage from which to launch surprise attacks on merchant vessels or passenger ships. Any captain who attempted to pursue Blackbeard through the inlet's narrow channels would almost certainly lose the chase.

Piracy along the coast was condoned and may even have been encouraged by high Carolina officials, including Governor Eden. Smugglers provided a ready source of corn, tobacco, and other goods that were not easily obtained through legal channels. They were an accepted part of the local economy.

But if Blackbeard was arguably good for Carolina, his propensity to shortstop cargos and terrorize men of commerce was bad for Virginia and, therefore, for its governor, Alexander Spotswood. Spotswood gained support from a group of upstanding North Carolina planters, who quietly urged him to put an end to Blackbeard's treachery. (Among those complaining was our cartographer-surveyor Edward Moseley, who also served as speaker of the lower house of the Assembly.) Operating in secrecy to avoid any detection of his plan, Spotswood dispatched an agent to North Carolina to determine Blackbeard's location. Instead of using two heavy Royal Navy men-of-war for the Ocracoke expedition, Spotswood wisely ordered two shallow-draft sloops to be outfitted with sixty men and placed Lieutenant Maynard in charge.

If Blackbeard was not too worried as Maynard's sloops approached him that Thursday night in November, we must suppose Lieutenant Maynard could scarcely sleep a wink. His position was unenviable. He would have to anchor on the Middle Ground between South and North Breakers: between the shoal stretching out to sea and the edge of the beach itself. Even so, he was in rough water. He would need to depend on the wind to hold him off the shoals, which lay north, south, and west of him. His two boats had to be prevented from slamming in the night, or fouling their cables. And the autumn wind is notoriously shifty.

We know the tide turned and was rising by about seven o'clock the next morning: Maynard was to use it to advantage and record this fact. So on that wearying evening it must have been rising, too, well into the dark of night. The rising tide would tend to carry the sloops into the wind and into the inlet—and onto the shoals. The fact that Maynard's sloops survived the night anchored off attests to a good, steady, prevailing southwesterly wind.

Still, many of Maynard's crew must have been up all night

to keep safe in such a precarious spot, quite apart from fretting over whether Blackbeard would make an escape or—worse—an attack. No doubt Teach died wishing he had tried one or the other.

By midnight the tide began to fall, ebbing with mean current out the inlet. This shift eased Maynard's plight somewhat, but only as long as the wind held westerly. Beginning in September, periodic shifts in the wind to north and northeast begin to override the prevailing southwesterlies. High pressure builds over Canada and the midcontinent north and west of the Banks. These systems produce an outflow of cool, dry air as they drift eastward and, eventually, out to sea. Storms, whether tropical hurricanes or frontal northeasters, will alter the autumn wind rose toward strong, seemingly prevailing northers, even though the lighter westerlies still blow most of the time. Fortunately, no such storm caught Lieutenant Maynard.

On slack tide of early morning Captain Teach slipped his anchor and let the current carry his sloop *Adventure* south toward the Point. His letter from his ally, the governor's secretary, Toby Knight, was veiled and vague. Ocracoke was as safe a place as he could hope to be, but if the governor of Virginia, Mr. Spotswood, would make trouble for him, Teach might have to fight. His crew was down to twenty-three (not counting himself) and some of those—cooks and boys—had not seen a pitched battle in their lives. So he would know who came in these two sloops riding off his inlet.

Maynard had to see the maneuver. One can imagine how he and his crew might have responded:

Ahoy, Mr. Hyde, don't be deceived by bare poles, yonder sloop is moving.

Aye, Sir, coming down to meet us she is.

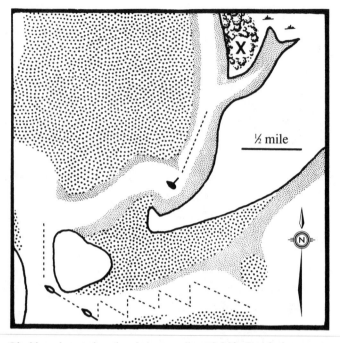

Blackbeard upped anchor but not sail, and drifted with the current down channel to meet Maynard's rowboat. Maynard and Hyde tacked their sloops through the Inlet, following their oarsmen.

I'm putting a boat and two men over to sound the channel, Mr. Hyde. Up your jib and bring Ranger *behind me. The tide slacks and the breeze holds light. We'll tack in past the Point.*

But, Sir, why not put men to the sweeps now? Tacking's a slow business.

No, Mr. Hyde. Every man at battle station, ready to fight. Those are no fisherman's poles across those sands. Once clear of the Point we can haul main sheets and run down on him.

But clearing Ocracoke Point that morning was no simple feat tacking under light sail. Maynard's sloop and Hyde's little *Ranger* grounded several times in the passage. Blackbeard held fast and waited.

Howard Pyle's vibrant reconstruction of the Battle of Ocracoke (*Book of Pirates,* 1921) is admittedly a blend of fiction and fancy. Much of it is unacceptable, but some of his points strike home. First, he has the small boat Maynard sent ahead sounding, under sail, like a dinghy. He correctly gives the distance as four to five miles from anchorage outside the inlet to Teach's Hole. Of course, Teach closed that distance by coming down the channel. When Maynard finally rounded the point and could run before the wind, the gap was still more than a mile.

Pyle's tale of the dinghy making two trips, the first to a "village" with "settlers," and a "wharf"—in search of an inlet pilot—is not credible. No village existed at Ocracoke before about 1735; there were probably no settlers or other long-term residents, and anything more than a crude dock and little ballast stone groin for a landing would be unwarranted. Blackbeard's hideout certainly featured a tent, and may have consisted of little else.

Still, Pyle's view of the timing of events is probably right. Bumping and grinding all the way, Maynard and *Ranger* (under Hyde) took hours to round the Point. Full high tide must have been about midday, and it ultimately saved the day for Maynard. The tide must have gone slack and begun to fall by about one o'clock. Though he did not yet know it, Blackbeard was doomed.

Still, he came very close to routing Maynard. Had he lured Maynard's sloops onto the shoal on a falling tide, history would have been different indeed. He let the dinghy come within a quarter mile: 1,500 feet. At that distance, across water, with only a light breeze, he could shout and be heard.

Perhaps he should have let the dinghy sail closer. He should have played for more time. But Blackbeard's crew was nervous and trigger-happy, and they fired their muskets at the dinghy. It was too far away. Their fire merely established hostile intent and sent the dinghy tacking back toward Maynard's sloop.

That impatience may have been Blackbeard's undoing. Running before the wind, the dinghy was well out in the channel. If it got too close before it came about, its first starboard reach would run it into the shoal. It was critical to Blackbeard's plan that Maynard not find the position of the shoal. So Blackbeard dared not hold fire for long.

No matter: the next sequence of events proved critical. With the dinghy back in tow, Maynard and *Ranger* set full sails and ran straight down on *Adventure*. That is when Blackbeard began his best work of the day.

Haul sheets, you maggots! Give me full sail now!

The pirate took the helm and ran before the wind, too, straight toward the beach. His quartermaster, Thomas Miller, accosted Blackbeard in a panic, but was shoved away. Then Miller saw the plan, the trap. Maynard knew nothing of the shoal that lay to his north, port side. Blackbeard was determined that he run aground on it.

Before piling on the beach, Blackbeard nimbly swung to the port reach. He did not need to come about. Noting his change of course, Maynard must have reckoned Blackbeard was now heading straight out into the sound, fleeing as the craven coward Maynard suspected all pirates truly were. But Blackbeard was not running. As Maynard swung port to cut him off, Blackbeard too came farther port, cleaving close to the wind, almost on collision course with the point of the shoal—and Lieutenant Maynard as well.

Blackbeard had eight deck guns, small cannons. Maynard

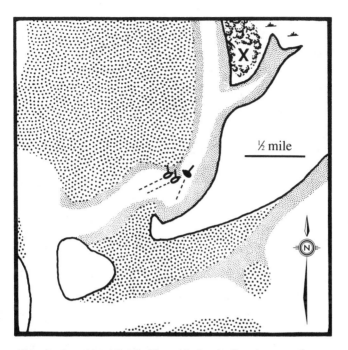

After the first skirmish, Blackbeard hoisted sail and ran north-northeast up the channel. In hot pursuit, Maynard and Hyde ran northeast—just as Blackbeard hoped—and thus ran aground. Blackbeard came about as Maynard and Hyde were forced to jibe, presenting their broadsides to Blackbeard's cannons.

and Hyde were equipped with none, but they had twice the number of men at arms. Maynard cannot have known of Blackbeard's cannons, but he certainly should have. This was his great failure of intelligence. In those early years of the eighteenth century, a pirate ship would outfit herself with four- to six-pounders, mounted on wheeled carriages. To hole a hull and sink a ship, balls were fired at very close range: forty or fifty feet. The trick was to depress the trajectory of the balls: literally to fire down from your own deck to the opponent's water line. To accomplish this one used a wooden wedge, called a quoin, which was driven between the breech (back end) of the cannon and the toe or butt (wheelless posterior extension) of the carriage. To elevate the barrel, one pulled back the quoin.

A pirate, of course, would not want to sink ships. He probably did not even stock cannon balls. He wanted to board boats safely afloat, plunder them, and then—if it struck his fancy—burn them. A pirate stocked *langrage*. That gives much greater range, up to 200 feet. One fires it high up. It falls down.

We know and use the expression, "everything but the kitchen sink." But for first-quality langrage, no such exclusion is required. The kitchen sink will make lovely langrage, once smashed into chunks and jagged fragments. So will horseshoes, nails, bolts, hunks of glass, even rocks, and especially chains. One can chain or tie good langrage together and sack it up in old sailcloth. Thus pirates were among the first to recycle their rubbish, especially the sharp, hard stuff that cuts, clobbers, and kills. Blackbeard, a prince among recyclers, was about to complete the cycle of his finest junk.

Maynard and Hyde were reaching smartly, closing rapidly on *Adventure*, and, though they did not know it, heading directly toward the shoal.

Shoal, Sir! Shoal dead ahead! Lord God, shoal to port, too, Sir! Coming up too fast, Sir. That's ground you could plow!

Starboard! Starboard! Haul the boom inboard. Fast, men, fast! Get her inboard and drop the sails!

Then the sudden, lurching stop: hard aground. Dead in the water.

What now? There might be minutes left to full flood tide, then a little slack, but soon—all too soon—begins the ebb.

To the sweeps, men! You've got to pull us off!

And they did. With the last of the rising water, with the westerly wind freshening just a little and helping push them east, toward the channel, half of them would get off alive and fairly well. Half, including Mr. Hyde of *Ranger* and both his top officers, would die or be badly wounded on that shoal.

Blackbeard was closing fast. The freshening breeze gave him speed, too. Speed to bring him in range before Maynard got off the shoal, but speed to precipitate the battle a fraction too soon to catch the falling tide.

At 400 feet: *Damn you for villains, who are you? And whence come you?*

Maynard's reply: *You may see by our colors we are no pirates!*

Blackbeard: *Send your boat on board so that I might see who you are.*

One can imagine Maynard's dinghy, now with men at its oars, trying to pull Maynard off the bar.

At 250 feet: *I cannot spare my boat, but I will come aboard you as soon as I can with my sloop!* A brazen declaration of war: the intention to board by force.

Maynard was close enough now to look down the barrels of Blackbeard's cannons. Suddenly Blackbeard dropped sails and came about. To anyone on shore, or the men on Maynard's sloop and *Ranger*, it must have looked as though Blackbeard ran aground, too. The pirate came about sharply and, held into the wind, *Adventure* stopped.

Blackbeard: *Pull the quoins! Touch fire to my guns!*

The range was perfect, the targets dead in the water. Blackbeard's deadly langrage fell with a vengeance. "This single broadside of eight cannons was devastating," wrote Robert Lee. The carnage inflicted by langrage at 150 feet is terrible. "With a single broadside . . . Blackbeard had reduced the attacking force to half its original size."

At this juncture we must part company with those chroniclers who contend that Blackbeard's ship went aground before his cannons were fired. Even if *Adventure* had grounded prior to firing (which we doubt), the recoil would serve only to push her off, back into the narrow channel. Here Blackbeard committed his fatal mistake. He could see the death and destruction he had wrought; he evidently believed the battle all but won. He did not hoist sail and run up the channel—the prudent course of action while reloading his cannons. He may not have bothered to reload at all, for there were no living men visible on the royal decks.

As the tide slacked, the outflow current took over, easing *Adventure* southward, down channel. The freshened wind pushed her hull east, toward the Ocracoke sands. Blackbeard was so intent on studying (or toasting?) his apparent victory that he failed to watch his stern. It was across the narrow channel that he went aground, a thousand feet southeast of Maynard's struggle on the shoals.

But Maynard's sails had been down when the langrage struck. They were cut up, of course, and the rigging much damaged. But not hopelessly so, as they probably would have been had his sloop been under sail. The men in the dinghy survived, as did some at the onboard sweeps, low to the deck and partially sheltered by the rail and cabin.

Maynard again: *Get below men. Leave me two—you, helmsman, and you, Butler: we'll get up the jib. She's afloat now. Stave barrels, lads; pitch over ballast and all we can live wi'out, for it's living a bit longer we must try to do!*

Blackbeard was out of cannon range now, so Maynard could scurry about, give aid and consolation to the wounded, and make sure his able-bodied were armed but hidden—below decks, out of the way of lethal langrage.

Meanwhile, *Adventure*'s stern grounded on the steep bank and her keel cut deep in the mud. Blackbeard was aground by the stern; falling tide, wind, and current conspired to keep him there. But he was not worried. His men were hard at work loading more bits of finer rubbish in bottles, and fitting these with powder and oil-soaked fuses. This pirate used grenades—which it is said he invented—to fine effect.

With only his jib pulling, Maynard closed the thousand feet across the channel. The second sloop, *Ranger*, captain and crew mangled, lay still aground. Maynard's courage certainly can never be questioned. He may at first have tackled Blackbeard believing all pirates doltish villains and cowards; but he had been grossly outwitted and out-gunned that day, while his adversary was inconvenienced but as yet unharmed.

As Maynard approached, Blackbeard and his hale crew could fairly swagger, readying grappling hooks and lines and preparing to touch fire to their grenade fuses. As Maynard's sloop drew close, Blackbeard made his second bad judgment of the day: he assumed all but three or four of Maynard's crew were dead or mortally wounded.

Grapples flew and caught; grenades caught and flew. The smoke obscured the deck of Maynard's sloop as the pirates swarmed aboard. But Maynard's men swarmed too: up out of the holds that had hidden them, as well as saved them from the impact of the grenades.

The grisly gore and gruesome details of that battle have been well-described. It is said that Blackbeard absorbed not fewer than five pistol balls at close range, and took twenty cutlass chops—one nearly severing his neck—before he finally dropped. A dozen of the royal crew were killed and roughly twice that number wounded. Numbers vary, but counting Blackbeard himself, at least ten pirates were killed and nine wounded. After the subsequent executions of the captured pirates, the toll of the dead would reach twenty-three.

Maynard finished chopping off Blackbeard's head and hung the souvenir from his bowsprit. As for his carcass's legendary swim three times around the sloop, we must dispute the story. With the falling tide, there was insufficient water by the stern for swimming of any kind. And as for the tale that the pirate's skull was plated with silver and has been used as a drinking cup by undergraduates at the University of Virginia, we have not been able to verify it. What can be asserted is that the ebb tide at Ocracoke that day ended a high-water mark of the Golden Age of Piracy, which by 1725 would be over.

Did Charles Johnson, later to become a captain, witness or participate in this battle? Was he a boy left at Blackbeard's camp, on shore? Or one of Maynard's crew? Not likely the latter, or we would know more of him. Captain Johnson wrote six years later, in 1724, very much as a narrator of sights seen at first hand. We give him the last word: "Here was an end of that couragious Brute, who might have passed in the World for a Heroe, had he been employed in a good Cause. . . ."

Chapter Five
Woods

The forest is a peculiar organism of unlimited
kindness and benevolence that makes no demands
for its sustenance and extends generously the
products of its life's activity; it provides
protection to all beings, offering shade
even to the axeman who destroys it.

Buddha, as quoted by
Huntington Cairns, 1973

One thing is certain: a few days of bulldozing is
enough to destroy the vegetation produced in
centuries, and we may lose such
prime vegetation forever.

Au Shu-fun, 1974

F RIENDS CALLED HIM Shu-fun Au, but in the construction of his native language, Cantonese, one's surname
comes first, just as it does in the West when we are serious
about being found—in the telephone book, for example. Dr.
Au came out of South China from Hong Kong, where he
was born in 1938. He earned his master's and doctorate at
Duke. He was killed in an accident near Kinston, North
Carolina, on October 17, 1969. His posthumously published doctoral dissertation is a monument in coastal research
and a testament to great promise cut short. His beleaguered
remnant maritime forest of western Shackleford is both an
island and an anchor in the ecology of the Outer Banks.

The three largest, most diverse, and ecologically best-
developed maritime forests left on the Outer Banks are
Shackleford at the farthest southwest terminus, Buxton
Woods just inland of Cape Point, Hatteras Island, and Nags
Head Woods to the north. They are of disparate ages, and
therein lies a key to the unique significance of each. All three
have suffered greatly at the hand of man—and the mouths
and hooves of man's livestock—and none is truly safe even
today.

Shackleford, Au's forest, is by far the youngest. Land in
the present area occupied by this woodland of loblolly pine
and live oak is less than 2,000 years old. Records attest that a
mere 200 years ago the woods of Shackleford were extensive
and broad at the eastern end of the Bank, attached to Cape
Lookout then as Buxton Woods is still attached to Cape
Point. At the time, this was the widest and oldest portion of
Shackleford Bank. Today it is all but eroded away. Dunes
flank its southern, seaward face, but sand flats or swales and

small marsh islets in Back Sound are all that now signal its former width. The west-end woods are the youngest originally present on Shackleford. Au found no live oaks more than seventy years old there.

Au's young woods on Shackleford are the safest on the Banks today. When Au was working there, the situation looked bleak. His writing is punctuated with his concern. He knew the Park Service planned to acquire Shackleford, but remained apprehensive. He cited literature documenting the trend to turn our national parks into "giant amusement parks and picnic grounds." The tendency to do just that, fortunately resisted when the Park Service did acquire Shackleford in 1976, has not entirely faded from government planning.

In Au's time feral livestock—goats, sheep, and horses—were destroying the vegetation of Shackleford. These exotic creatures, brought to America and the Banks by Europeans, eat the acorns and seedlings of live oaks. They reduce—and in large numbers actually destroy—the ability of the forest to recover from natural disruptions like hurricanes and lightning-strike fires.

But the effects of livestock are more insidious in indirect form. They graze the dune grass, especially sea oats, right down to the roots. This denies the plants their ability to take in sunlight and convert carbon dioxide to nutritive carbohydrates. Worse, perhaps, the hooves of livestock cut into the sand and sever the lattice of subsurface roots. Dune grasses are really vast, extended, single plants—many tufts interconnected by a web of roots. These roots run largely just below the surface, but occasionally, here and there, penetrate deep down to the "freshwater lens," a convex layer of groundwater. Cutting an individual tuft out of the network of roots dooms it in the summer sun.

The demise of the grasses frees the sand to move. Moving, drifting sand consumes the shrub zone and engulfs the forest. Some of that consumption is natural, and the vegetation can keep up with it, or recover from it. But at the novel rates produced by the grazing and trampling of foreign livestock, no plant community can hope to keep up for long.

Fortunately, among the first moves the Park Service made on Shackleford was to eliminate the livestock. The tradition of using federal land for free or cheap grazing of privately owned animals runs long and deep in America. The Park Service made Shackleford a rare exception, an anomaly, in the general course of events. So Shackleford today is recovering. Au's seventy-year-old trees are approaching their century. Au's work was not in vain.

The fauna of Shackleford has not been thoroughly studied. Dr. William Engels, who discovered the distinctive reptiles of the Intercapes Zone before World War II, found southern, banded water snakes (*Natrix fasciata*) there. This species is not found elsewhere on the Banks, but it is common along the mainland north to Albemarle Sound. Engels also found the greenish ratsnake (*Elaphe obsoleta*), a species common on the mainland, but which is present on the Banks only far to the north, from Nags Head to Buxton Woods. Otherwise, the herpetofauna seems so far unremarkable, but much of it may have been lost.

Numi Goodyear (then Spitzer) conducted a mammal survey published in 1977. The only typically mainland species she found—not also found in the Intercapes Zone—was the mole (*Scalopus aquaticus*). Moles are common in Buxton and Nags Head woods, and occur here and there along the Banks between these two areas. They are absent from the Intercapes Zone between Buxton Woods and Shackleford. Apparently they simply cannot survive the level of aridity and frequency of overwash encountered in the Zone.

The moles Goodyear unearthed at Shackleford were a beautiful, lustrous, golden color—a variant known elsewhere along the southern coasts. Those we have seen from Buxton Woods are ordinary, slatey or sooty.

Nags Head Woods is relatively ancient. Its flora and fauna are rich compared to that seen in the much newer forests at Shackleford and Buxton. Many isolated relict species live here, which attests not only to great age, but to different climates and former ecologies as well.

Nags Head Woods probably first developed as a dune field on the edge of the falling sea at the end of the Sangamon Interglacial. During this warm period, high-stand sea level was about 20 meters (or 70 feet) above its present position, and the coastline lay far inland. Nags Head was underwater, of course. As new glaciers formed, sea level dropped and the dune field developed, something like 60,000 to 80,000 years ago. Sea level went right on dropping, until at the low point in the Würm Glaciation—50,000 to 20,000 years ago—the coastline was at least 100 meters below present sea level, and far eastward, at the edge of the continental shelf.

During the Würm Glaciation, the last to date in earth's history, Nags Head Woods was a high point on a vast, broad, continental plain. The climate was cooler than to-day's, although glacial ice lay far to north, never closer than Long Island, New York.

Species characteristic of the preglacial north extended much farther south during the Würm. Some of these today reach the southern limits of their ranges in and around Nags Head Woods. Examples include wooly beach heather (*Hudsonia tomentosa*) and bayberry (*Myrica pennsylvanica*), so characteristic of Long Island and Cape Cod dune habitats. The

American toad (*Bufo americanus*) survives here today, a disjunct outlyer this far south on the coast, but characteristic of the Appalachian uplands and the north—all the way to Labrador and Hudson Bay.

Some 5,000 to 4,000 years ago, the climate warmed dramatically. It was notably warmer than it is today, although sea level—rising for about 15,000 years by that time—was still lower than it is now. Southern species moved northward along a coastal plain that is now largely submerged. Then the climate cooled somewhat, but sea level continued upward. Some southern species, whose ranges during the warm period (called the Hypsithermal Maximum by scientists) had been continuous, survive as remote disjunct outlyers—far north of the main ranges of their species—in the ecological island of Nags Head Woods.

Examples include the palmetto, *Sabal minor*, such a typical indicator species of woods and hammocks from Buxton southward, and water penny, *Hydrocotyle bonariensis*. Most remarkable, perhaps, are several Deep South reptiles now relict in these woods. The yellow-lipped snake (*Rhadinea flavilata*), black swamp snake (*Seminatrix pygaea*), and chicken turtle (*Dierochelys reticularia*) are examples.

Today, a walk around Nags Head Woods—partially set aside as a preserve under the auspices of The Nature Conservancy—reveals scenes that could be duplicated in Great Smoky Mountains National Park, on the one hand, and Croatan National Forest, far south of the Neuse River, on the other. And these incredible juxtapositions are crammed into a mere 1,400 acres—as yet incompletely preserved for future generations.

By far the largest remnant maritime forest on the Outer Banks is Buxton Woods. These woods in many ways constitute the most distinctive region of the whole Outer Banks.

As we have seen, this land mass is about four miles wide north to south; it spreads almost due west from the Atlantic just north of Cape Point for about eight miles. In all, the Buxton highlands add up to more than thirty square miles, or nearly 8,000 hectares, approaching 20,000 acres. About half of this land area was once covered by maritime forest, but only 3,000 acres of forest remain today.

Buxton Woods is the largest and most isolated piece of maritime forest remaining in North Carolina. It is also one of the most endangered. A recent survey commissioned by the state concluded that if immediate steps were not taken to preserve these rare woods, they and all the other dwindling patches of maritime forest along the coast would vanish within ten years. Thus would man destroy what nature has taken thousands of years to build.

The broad, dune-corrugated expanse of the Buxton highlands is noted for a unique ecological feature: a huge freshwater lens. The freshwater lens is a wonderful feature of sand land. It develops from rainwater trapped within the porous sand. Other kinds of land, like limestone, can trap and hold rainwater, too, and develop freshwater lenses; but never to the volume that sand does. The freshwater is called a *lens* because of its shape. In side view, the freshwater looks like a glass lens, convex above and below. This shape is derived from the fact that freshwater is less dense—lighter—at a given temperature than saltwater. Salt molecules in solution tend to fill the spaces between water molecules; saltwater simply possesses more molecules of stuff—water plus salt—per unit volume than freshwater (at a given temperature; temperature changes alter densities).

Virtually all terrestrial life depends directly or indirectly on freshwater. If a freshwater lens did not float on the seawater substrate within the sands of the Banks, most terrestrial life forms could not exist here. The freshwater lens within the Buxton highlands supports the woods and the freshwater swamps and swales. These sustain the birds—especially thousands of migrants each year—and all the other wildlife of this remote, wooded island out at sea.

The fauna and flora of Buxton Woods—except for those members who can fly—all had to disperse to these woods either across the sounds or along the Banks, and they had to make the trip in less than 5,000 years. While Nags Head Woods has been in place since the beginning of the Würm Glaciation, at least 50,000 years, the site of Buxton Woods was awash from at least 10,000 to 5,000 years ago. Buxton Woods has been allocated a tenth of the time Nags Head Woods has had to develop a maritime forest ecosystem.

Today, however, many of the forest plants and animals of Buxton Woods are isolated disjuncts, too. Both fauna and flora are under study. Susan Bratton and Kathryn Davidson have researched in detail the history of the vegetation in Buxton Woods. Davidson's vegetation map of the area, drawn up in 1984, was published by the Park Service in 1988.

Alvin Braswell, of the North Carolina State Museum, and David Webster, at the University of North Carolina at Wilmington, have studied the herpetofauna and mammals, respectively. Braswell found seven species of amphibians (six of them frogs or toads) and nineteen species of reptiles for a total of twenty-six species. This compares to forty-six for Nags Head Woods. Braswell learned that most—61 percent—of the Nags Head herpetofauna was characteristically mainland, while the remaining minority—39 percent—consisted of widespread barrier island forms.

At Buxton Woods, Braswell found just about the opposite: only 35 percent are mainland types, while the majority—65 percent—are typical of barrier islands. Down in the Intercapes Zone, at Ocracoke, only 10 percent of the dramatically reduced fauna could be thought of as mainland types. So far, isolation of the continental reptiles and amphibians at Buxton Woods seems not to have produced notable evolutionary departures comparable to the two snakes of the Intercapes Zone, but close study of some of these insular gene pools will no doubt reveal new information about time of dispersal and degree of genetic differentiation.

David Webster and his students have compared mammal faunas of several of the Carolina barriers. Webster's work on Buxton Woods is a fourteen-page typescript dated March 1988, reported to the Coastal Resources Commission. Since then we have communicated directly and received updates as his work progresses. Some thirteen terrestrial mammals and two species of bats are now well-documented for Buxton Woods. Opossum, mole, gray squirrel, white-footed mouse, and deer appear to be isolated disjuncts. At least the mouse seems to be unique, novel, an evolutionary departure.

We first met the mouse in 1971 when, using then-student Numi (Spitzer) Goodyear, we were trying to determine what kingsnakes could find to eat. Goodyear caught a fair sample of these pretty mice and deposited them in the collection of the University of Massachusetts, Amherst. We immediately thought them different from other native mice of the genus *Peromyscus.*

The Buxton Woods mouse is quite large—much larger than the mainland white-footed mouse, *Peromyscus leucopus.* This feature seems to have led some previous workers to assume it is a representative of the cotton mouse, *Peromyscus gossypinus.* The two species are extremely similar. Indeed, where they occur together, size is the best distinguishing feature.

The white-footed mouse is a northern form, occurring along the mainland coast as far south as the Neuse River, which empties into Core Sound. Most are richly colored in warm shades of brown and fawn with near-black guard hairs especially dense in a strip down their backs. The underparts are snow- to cream-white, very distinctly and sharply set off. A small, pallid form occurs at Virginia Beach. The cotton mouse, a southern species generally more common along the mainland coastal plain, occurs northward only to south-eastern Virginia. It has a more orange or yellowish cast—*ochraceous,* scientists say—and is more overcast with gray, including a grayer whitish belly.

The facial features—difficult to quantify—also seem different between the two species. The white-foot has a more convex profile and a blockier head. The cotton mouse bears a sharp-nosed look. We thought the Buxton Woods mouse was a white-foot derivative, not related to the cotton mouse. Others disagreed, and for a time both species were listed as denizens of the Woods. Only one is really there.

The final word has not been heard, but Webster inclines toward the white-foot camp. In addition to large size, this possible new species is paler than the typical mainland white-foot, but much more colorful—a sort of café au lait on top—than the Virginia Beach variety. We hope Webster and company will resolve these various issues soon, so that Buxton Woods may have its own unique animal.

WALKING ALONG THE beach, combing for shells and dodging the tide, one does not immediately think of woods. If it is the cool shade of a tree one longs for, the edge of the Atlantic coast seems the last place to look. There are a few scraggly stands of redcedar or pine scattered here and there, but most woods worthy of the name are located well inland of the beach itself.

Strolls in the woods were certainly not what Shay and Kim Clanton had in mind when they moved to the tiny community of Frisco on Hatteras Island in 1985. Like many of the island's new arrivals, the young couple were drawn there by its remoteness from civilization and by its proximity to the sea. Shay, an artist and Virginia native, observed in the Banks' stark landscape inspiration for her paintings. Kim, a North Carolina native with eclectic interests from glass-making to commercial fishing, found in the Banks an outlet for his talents.

It was only after settling into their small, breezy home on the south edge of Frisco that Shay Clanton began to be fully aware of the Woods. She knew woods were there, of course. You cannot drive along Highway 12 from the outskirts of Frisco in the south through the town of Buxton to the north without noticing the Woods. It lies on both sides of the road, dark and impenetrable. Lots for houses and stores have been carved into its periphery along the way. But it does not yield its secrets easily; Hatteras natives know that the Woods can be infested with cottonmouth moccasins and mosquitos. Still, Shay Clanton began to immerse herself in the Woods. A path that led into it began almost at the doorstep of her house. Clanton started walking, and she quickly discovered the wild beauty and amazing diversity of Buxton Woods.

She followed a network of paths developed and used by islanders over many years. Along a parallel series of high relict dunes that represent ancient Hatteras shorelines, Clanton admired sandy, needle-covered forests filled with large pines and oaks. Between those ridges, she explored low-lying, freshwater ponds and marshes, called *sedges* by the natives. They were surrounded by palmettos, holly, even native dogwood. There, forty miles out to sea, the dogwood bloomed that spring, as it blooms so brightly every spring throughout North Carolina.

Shay Clanton saw that, and then she experienced perhaps the most astonishing sensation of all; that is, to stand in a cool, shady forest of towering pines in Buxton Woods, and to hear a sound so familiar that it is surely archetypal—the unmistakable gush of the ocean just beyond the forest's edge, where it yields to dunes and their hardy stands of yaupon and wax myrtle. Clanton could not see the ocean from her perch atop that ridge. She would not have known it was there, except for the crash of breakers on the beach. But it *was* there, and despite its salt and harsh winds, the ocean somehow tolerated a magnificent stand of forest only a seagull's swoop away, just behind the front lines of dune.

Besides the clouds of mosquitos, Shay Clanton's early walks summoned questions. How could a large forest manage to thrive so near the ocean? How did freshwater ponds bubble up within its confines? How did species of trees and plants from the mainland elope so far out to sea, with nothing but saltwater and sand on all sides for many miles? Clanton wondered, too, about the Buxton Woods aquifer that served as the source of all the fresh water for the towns of Buxton, Frisco, and Hatteras. Without that water, no human (and very few plants or animals of any stripe) could survive there for long.

It was in the midst of such musings that Clanton spotted the item in the local newspaper. It was an interview with the president of a group of investors—from on and off the island—who described their plans for a new golf course and housing complex. To Clanton's dismay, the development was to be built on 163 acres in the center of Buxton Woods. The article praised the developer for his energy and cited his desire to protect the environment.

Clanton was alarmed. She knew the area in which the golf course was to be built. It would effectively cut out the heart of the Woods. If the project were built as described, the Woods as she knew it—as a distinct ecosystem—would not long survive. Moreover, the virulent real-estate boom on the Outer Banks that had spawned the proposal would usher in other new threats to the Woods. The relative cushion the Woods afforded from sea and storms had already made it a prime target for new housing.

A quick check of local ordinances also revealed that Hatteras Island had no zoning laws. You could build whatever you wanted on whatever land you owned. The only real zoning was conferred by the elements, which saw to it that structures that failed to conform to the island's specifications were summarily flattened. But that rule did not apply as effectively to the Woods, whose high ridges offered some protection from winds and storms. It was possible to imagine a lush, green golf course nestled in the crook of Buxton Woods, the next Hilton Head resort waiting to be born. It was to be called the Cape Hatteras National Golf Course.

Shay Clanton did the only thing she knew to do. She wrote an acquaintance who worked in state government in Raleigh. In retrospect, the letter represents a modest turning point in the island's history. It was the first small shot in the battle, still continuing, to preserve Buxton Woods for future generations. The letter began:

I am writing to you to see if you can help me with a problem here on Hatteras Island. A group of developers here are planning a golf course and housing development on 163 acres of maritime forest known as the Buxton Woods. The area is the only heavily wooded area on Hatteras Island and it is not only a unique and beautiful forest, but it is vital to the water supply of the island. . . . A golf course will require the destruction of most of the trees, it would require a tremendous amount of water for daily maintenance, and it would create the potential for pollution from herbicides and pesticides. Surely the replacement of a large section of the maritime forest of a barrier island by a golf course represents a radical disturbance of the ecology that would be very destructive to the whole island.

The letter went on to ask if a study could be conducted to determine what the project's impact would be on the Woods. What Clanton did not know was that the state of North Carolina had begun to develop an interest in preserving coastal areas. By designating certain environmentally sensitive areas as so-called Areas of Environmental Concern (AECs), the North Carolina Coastal Resources Commission could impose controls on development.

Clanton's letter eventually wound up on the desk of Rich Shaw, a young program analyst with the state Division of Coastal Management. At first it appeared that the state could not intervene in the Buxton Woods matter, because maritime forests weren't included on the state's original list of AECs. The only AEC on Hatteras Island was the well field in Buxton Woods that provided water to neighboring towns. The rest of the Woods, except for the 900 acres protected as part of the Cape Hatteras National Seashore, was completely vulnerable to the whims of its owners.

With support from Shay Clanton and other island residents, the North Carolina chapter of the Sierra Club petitioned the Coastal Resources Commission to nominate portions of Buxton Woods for AEC designation. A commission staff study, backed up by a panel of scientists, confirmed what Clanton's walks had revealed: that the forest is a unique and fragile resource that could be easily destroyed by uncontrolled clearing and development. When one part of the forest canopy is removed, it exposes the underlying trees, shrubs, and flowers to harsh wind and salt spray. They cannot long survive the intrusion. In Buxton Woods, even more immediately critical was the potential threat such development posed to the island's underground water supply.

Meanwhile, Shay Clanton was hard at work organizing a group to monitor developments and lobby for zoning controls. She had talked idly with a friend and Hatteras native, Carol Anderson, about forming such a group, but the golf course proposal galvanized them. Together with Roy Johnson, a retired insurance broker who had moved to the island, Clanton and Anderson founded the Friends of Hatteras Island. The group's first meeting, with some twenty-five people on hand, was held in a tiny Buxton bookstore. The Friends would soon grow into a thriving organization more than 250 strong and serve as a model for other Outer Banks residents seeking to protect their environment.

The Friends' first task was to try to postpone the golf course development until the AEC designation could be considered. The Dare County planning board complied. Digging in the county's deed office, Roy Johnson unearthed a key piece of evidence. He found a map drawn up in the 1960s indicating that the Woods' well field extended into the area designated for the golf course. This was an additional area that the Cape Hatteras Water Association had apparently set aside for expansion in future years, as the island's population grew.

Johnson's discovery led to an appeal from the Friends of

Hatteras Island that the state's AEC designation be expanded to include the new area. The Coastal Resources Commission agreed. It informed the golf course investors that they would have to acquire development permits and have their building plans reviewed by state and federal agencies. They would also be required to show that no pesticides, fertilizers, or septic runoff from the project would endanger the well field.

Several months later, the commission agreed to extend the width of the well field farther into the golf course property. The project was abandoned. Instead, the owners agreed to sell the land to the state as part of a new heritage program to preserve valuable coastal property for research, education, and public use. In the fall of 1987, Governor Jim Martin announced the purchase of 162 acres of Buxton Woods for a price tag of $750,000. Instead of fairways and greens, there would continue to be pine and oak, hornbeam and holly—and an occasional moccasin.

As for the drive to confer AEC designation over much of Buxton Woods, heated debate ensued. Some residents saw the proposal as a threat to local control and to their pocketbooks. They also objected to the minimum lot sizes that would be imposed on building in the Woods. The Friends of Hatteras Island argued that to permit uncontrolled development would destroy the last large stand of maritime forest in North Carolina.

Initially, the Dare County Board of Commissioners opposed the AEC designation. But as negotiations continued and it became clear that the state was prepared to step in, the county proposed its own set of local regulations in which Buxton Woods would be designated as a Special Environmental District. Under this plan, lot sizes in the Woods could not be smaller than one acre, clearing of vegetation

would be controlled, and impact of development on wet-
lands and water supplies would be closely monitored.
Though the county's plan was less stringent than the state's,
the Coastal Resources Commission eventually accepted the
local proposal as a reasonable compromise. A significant
byproduct of the debate came when the county commissioners adopted the island's first zoning code.

While initially skeptical, the Dare County Board of Commissioners learned to respect the Friends. Like Robert Dolan
and Paul Godfrey before them, the Friends had compelling
facts on their side. The Buxton Woods compromise was an
example of environmental activism and responsible government working in tandem to protect a precious natural
resource. In a brief two years, Buxton Woods had moved
from no protection to significant protection. A good working relationship between county officials and the Friends had
been forged. The large tract of land designated for a golf
course and housing development had been purchased by the
state, and there was talk of further land acquisition. Reason
to celebrate? Of course. The long, weary hours that Shay
Clanton and her Friends had devoted to saving the Woods
had paid off handsomely, beyond their fondest hopes. But
the struggle was not over.

As we write, the future of much of Buxton Woods remains
in doubt. Petitions to designate larger sections of the Woods
as Areas of Environmental Concern stalled. A divided
Coastal Resources Commission concluded that outright purchase of the Woods, rather than stricter controls on its
development, was the only way to assure its preservation for
future generations. That is no doubt true; but time is short.
Where is the money to purchase the Woods to come from?

In 1989 the federal government provided a grant of $1.5
million to buy property within Buxton Woods. But to

receive any of the money, the state was obliged to match it. A year later North Carolina forfeited $281,000 of the federal money because it could not meet a deadline.

Still, by mid-1991, the state had acquired nearly 450 acres in the heart of the Woods, and more purchases were planned. It appeared that an increase in the state's land-transfer tax would make more funds available for purchase of threatened, environmentally sensitive land, including Buxton Woods. The Nature Conservancy had stepped in to speed up the cumbersome process of identifying available tracts of land. Most heartening, perhaps, was that local governments had begun to fill the vacuum created by the Coastal Resources Commission's hands-off approach. The towns of Kill Devil Hills and Pine Knoll Shores, and the village on Bald Head Island—all sites of threatened maritime woods—adopted controls on development using the Buxton Woods model.

But development in and around Buxton Woods has continued. The latest attraction is a miniature golf course built on a 1.8-acre tract along Highway 12; there would be golf played in Buxton Woods after all. More ominous, perhaps, the Cape Hatteras Water Association has resorted to a lottery to auction off its last water hookups to anxious landowners. But the association has asked the state's permission to drill nine new wells in Buxton Woods in order to pump another million gallons of water a day. The Friends of Hatteras Island fear the additional pumping will jeopardize the aquifer and spur new development. Thus might the very entity invoked to spare the Woods—the freshwater well field—contribute to its demise.

≈

EVERY BIT AS interesting as the woods themselves are their edges. Of the three maritime forests noted on the Banks, Buxton Woods has the most signal and remarkable edges. Buxton Woods marks a boundary not just in the arcane investigations of scientists, but in literature as well. Writing in 1966, Charlton Ogburn, Jr., noted it: "The winter beach in any case went no farther than Cape Hatteras. At ... Buxton a few oleanders were growing, a ... myrtle was heavily draped with Spanish moss and there was ... palmetto in the woods. This was the frontier of the Deep South."

Anyone can see The Line. It is certainly one of the clearest edges visible in ecology, evolutionary biology, and biogeography. It seems almost as abrupt and unequivocal as the line between land and sea. One can see it from an airplane or a satellite. It photographs beautifully. One can see it from the ground. Just stop along the road to Frisco Campground, walk uphill to get a vantage point, and look east and west: there it is—the edge of dark green shrubbery simply stopping at grass and sand.

North of The Line is Buxton Woods. South of The Line is the Intercapes Zone, extending from Cape Point of Hatteras Island southwest for seventy-six miles to Cape Lookout at the nether tip of Core Banks. Within Buxton Woods, and in patches on northward to the Banks' confluence with the mainland, are many species of flora and fauna not found in the Intercapes Zone: species of the mainland, often species of the north.

Signal indicators are loblolly pines (*Pinus taeda,* some of which have been planted in the Intercapes Zone), laurel oaks (*Quercus laurifolia*), hornbeam (*Carpinus caroliniana*), red maple (*Acer rubrum*), white-footed mice (*Peromyscus leucopus*), and timber rattlesnakes (*Crotalus horridus*). There are few other places on the planet where waves of fall migrant birds—species that breed as far north as trees grow—can be witnessed departing for the tropics in such abundance.

Charlton Ogburn's oleanders are introduced exotics—climatic indicators perhaps, but not native. His "Spanish moss," actually a twining bromeliad (*Tillandsia usneoides*), grows well up into Virginia. But his palmettos are native and signal. However, Ogburn saw them on the other side of the Woods, at the town of Buxton, and they are an abundant and important component of the woodland understory throughout Buxton Woods. With the palmetto, complexity is introduced: there is a second line, the northern edge of palmetto.

In floral studies of the Intercapes Zone, palmetto is often omitted, imparting the distinct impression that those growing in Buxton Woods are a disjunct, isolated colony. This is not the case. Palmettos are a standard component of the live oak–redcedar Intercapes hammocks, occurring at Hatteras (ten miles southwest of Cape Point), Ocracoke, Portsmouth, and Guthrie Hammock on Core Banks. Palmettos are characteristic of well-developed, old growth hammocks as far north as Buxton Woods. The problem is that settlers waged war on the Intercapes' hammocks, cutting the live oak for lumber and smoking fuel, and the pungent redcedar for nearly immortal constructions. Where these activities have been suspended, palmetto has made a comeback.

There are other lines, too. For example, those white-footed mice occur southwest in soundside thickets, often dominated by perfectly natural pines, as far as the northern side of Joe Saur Cove, at the edge of The Line. The white-footed mice bring with them the ratsnake, *Elaphe obsoleta*, found nowhere within the Intercapes Zone proper. The ratsnake feeds on woodland rodents, and these stop, not to occur again until one gets to Shackleford. Then the ratsnakes appear again. Thus, around The Line we see so clearly are really many lines, a great edge that is a region of many edges.

Ecologists have long been fascinated by the edge phenomenon. They gave edges the special name *ecotone* to celebrate their unusual properties. The edge of the sea is a classic case. Here we find species not present in either major habitat—on the land or in the deeper water—flanking this ecotone. Ghost crabs, sand crabs, and coquinas are prime examples. Often species that do utilize habitats on one or both sides of the ecotone are disproportionately abundant right along it. Gulls and terns are good examples.

Ecotones on land, such as The Line, have been relatively little studied, although the same phenomena of uniqueness and abundance have been long and often well-noted by ecologists. Some of our most important and influential wild species, like poison ivy and white-tailed deer, are classically ecotonal in distribution and abundance. Perhaps the most obvious ecotones on land are artificial, the products of human activity. Few things can be more clear-cut, in appearance as well as etiology, than the edge of agricultural cultivation at the woods—that ecotone most favored by those deer and resplendent poison ivy. Natural ecotones on land are often less conspicuous.

The Line between Buxton Woods and the Intercapes Zone is simply one of the most conspicuous natural, terrestrial ecotones one can find. Fortunately, quite a bit of it is owned by the National Park Service; but not all. To study this ecotone properly we need to keep all its constituent parts: all those lines and edges that tend to amalgamate and interweave around and along The Line we can so easily

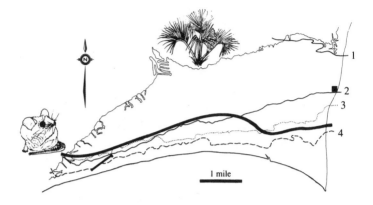

Ecological Lines at Buxton Woods. The heavy, sinuous black line is The Line photographed from Apollo 9, corrected for earth's curvature and angle of shot: 1) the northern limit of native palm, *Sabal minor,* sketched top center; 2) the southern limit in Buxton Woods of hornbeam, *Carpinus caroliniana,* a characteristic mainland hardwood also called blue beech; 3) the southern limit in Buxton Woods of loblolly pine, *Pinus taeda.* (Note that toward the west, pine and hornbeam lines cross and hornbeam actually extends farther south.); 4) the southern limit of native mouse range, an elegant, white-footed *Peromyscus* sketched far left. The black, angled line, lower left, is the runway of Billy Mitchell Airport. The black square, far right, is the Hatteras Lighthouse. Sources are various, including our own field notes, photographs, and specimens—and Kathryn Dawson, "Vegetation Map of Buxton Woods, North Carolina," National Park Service, 1988.

perceive. We need those mouse-supporting thickets down to Joe Saur Cove. We need the pine "islands" in the soundside marshes between Frisco and Buxton.

And, most especially, we need Buxton Woods. Many of the species—such as timber rattlesnake—that survive in Buxton Woods are genuinely rare now. Loss of just a little more habitat could push them over the brink to extinction.

The Line at Buxton Woods, from a photograph taken by Wenhua Lu, May 22, 1989, looking west from east of Billy Mitchell field toward the Frisco water tower.

Populations reach a lower limit below which they cannot withstand the vicissitudes of life. That is what happened to the heath hen of New England. Outright human predation reduced this highly edible species to one population on Martha's Vineyard. Then a sequence of accidents—brush fire, cat kills, disease—further reduced them until, on March 11, 1932, only one was left. No one knows just when he finally died and his species became extinct forever.

The rattlesnakes of Buxton Woods present a wonderful opportunity for genetic research. They combine characteristics of the northern, nominate timber rattler (*Crotalus horridus horridus*) and the southern canebrake rattler (*Crotalus horridus atricaudatus*). Rattlers from the adjacent mainland, as far north as Virginia's Great Dismal Swamp, are typical southern canebrakes. The Buxton Woods population is a widely isolated disjunct; the next closest population is also insular, in Nags Head Woods.

South of The Line is the Intercapes Zone, between Cape

Point of Hatteras and Cape Lookout of Core. It is a sort of weird world of deprivation. Mainland influence has crept down and across the Banks to reach for the Intercapes Zone in a pincerlike movement. Mainland fauna and flora have reached Shackleford and Buxton Woods en masse, but few members have straggled into the Intercapes Zone. Here, in a hotter, drier, more windblown, overwashed, hurricane-struck region, the initially deprived elements of an already reduced mainland biota were culled and directed by natural selection. Peculiar ecosystems, whose very peculiarity stemmed from the reduced numbers of the member species, brought new faunal and floral relationships.

Compared to mainland Carolina, there are hardly any snakes here at all. One might as well say compared to South America, there are hardly any turtles in the Galapagos. It is what happened to the ones that survived there that is amazing. In less than 5,000 years the Intercapes Zone forged two unique snakes called *autochthons:* they originated in evolutionary history in the Intercapes Zone.

One, a very dark and somber salt marsh snake, *Natrix sipedon williamengelsi,* is named for Dr. William Engels, professor at the University of North Carolina after World War II, who discovered the form just prior to shipping out as a lieutenant. Dr. Engels' salt marsh snake certainly derived from individuals that moved down the Banks from the north. It has invaded westward, across the sounds, into the ranges of its southern relatives. Its ecology and behavior seem comparable to other salt marsh snakes, both north and south.

The second, a kingsnake, has remained endemic: found only here. Living in ecological deprivation (compared to mainland kingsnakes), it has been winnowed by natural selection to fit a narrow and remarkable niche—and in a very short time. Few and far between are opportunities to study rapid, adaptive, evolutionary change such as is demonstrated here. The evolutionary biologist seeks to understand the process of genetically controlled adaptive change. Indeed, few subjects could ultimately prove more compelling for investigative scientists.

We are a species that evolved rather quickly and changed radically in physical characteristics from its ancestral stock. We cannot, however, either completely sever our hereditary ties to our apelike ancestors or stand still in our evolutionary state. We will continue to change, to adapt, or die. Conscious thought and decisionmaking ability set us apart from other animals, but have not yet provided us with the wisdom to plot our own evolution. If we are to develop that wisdom we will have to understand evolutionary processes, beginning with their fundamental biochemistry, genetics, and reproductive strategies. Examples of rapid evolution fairly cry out for investigation in these respects.

In the bizarre case of the Intercapes kingsnake, we scientists stand on an investigative threshold. We are ready to begin learning more, but it is a threshold we had to reach. It was that climb that brought us to the Banks in the first place, and that introduced us to the roots of the connections we celebrate in these pages.

Chapter Six

Trilogy

As is well known, the sandy, almost waterless, islands
off the Carolina coast have a very limited and naturally somewhat highly specialized fauna and but relatively few
forms have been able to adapt themselves to this
highly rigorous environment. The first of these
novelties . . . is . . . a King Snake.

Thomas Barbour and
William Engels, 1942

And I wished he would come back, my snake.
For he seemed to me like a king,
Like a king in exile, uncrowned in the underworld,
Now due to be crowned again.

D. H. Lawrence, 1928

A PLAUSIBLE DESCRIPTION of Hell: mid-June, 1971, camped out on a sandbar thirty miles at sea. Mud, mosquitos, greenhead flies, deerflies, and snakes. All anathema to this motley human crew, except for one—the snakes.

There were six of us. James Lazell had been in residence since June 12. John Alexander had just arrived. With Lazell were four secondary students. The three boys, Paul, Ned, and Mark, sixteen or seventeen, were animal catchers whose captures would figure heavily in the documentation of wildlife on the Outer Banks. The fourth was a girl, fourteen: Numi Catherine Spitzer.

The darkness thickened. The greenheads and deerflies, mercifully, were gone for the night. Mud-caked sneakers and trousers were hung out to dry. A cooling plunge in the ocean had refreshed the group. We were beyond all but the whine of mosquitos, inside a screened tent fly. Thick cobia steaks—back then regarded as junk fish by the Bankers—dripped flashing oil through the rustic charcoal grill. Shrimp steamed on the gas stove.

"Tell the story," said the girl. "Tell Mr. Alexander why we are here."

And so Lazell told us the tale. He told how he and Alexander had met in the fall of 1957, 2,000 feet above sea level, on the edge of the Cumberland Plateau, near Sewanee, Tennessee—Alexander in his last, eighth-grade year at the public school; Lazell in his first, freshman year at the University of the South. Hunting squirrels, Lazell was camped out when Alexander stumbled across him. They became fast friends, hunters of animals, stories, facts. . . .

Lazell explained, too, what it is that drives some people—a

very few, a select group, a group some might call peculiar—to
hunt snakes for a living. Herpetologists, they're called, in
the jargon of science.

Most people recoil when they think of stumbling across or
worse—handling—a snake. They will touch or keep as pets a
dog, a cat, a gerbil, a parakeet, or even a ferret. They think
nothing of hooking a worm or a cricket to go fishing. But
the thought of gathering up the most harmless of garden
snakes in their hands is disgusting to huge masses of otherwise sensible people, including many who enjoy the outdoors.

Part of this phobia is a consequence of the fact that some
snakes, though only a few, are poisonous. And some harmless snakes not yet accustomed to their human handlers will
bite the hand that grabs them. (So, of course, will a dog, cat,
gerbil, or parakeet, but apparently that's different.) Still, the
widespread phobia of snakes exists in a category by itself,
a monument as much to superstition and lack of knowledge as anything else. This innate dread may also spring
from the serpent's mythological association with original
sin; did not the snake tempt Adam and Eve to taste the
forbidden fruit?

But these irrational fears do not erase the remarkable
compulsion that snakes produce in us. You can see this at
any zoo. More people flock to the snake exhibits than to any
others except apes and monkeys: our kin so close we visit
them on Sunday afternoons. We do not visit snakes. We
stare at them. Crowds stare at snakes who, lidless, ever
aware, stare back. There is a fascination, an awe.

For some—the instinctive hunters—this fascination mani-
fests itself in the compulsion to search. Snake-hunters are
born, not made. Kids—some kids, funny kids—know where
the snakes are. They know by some primordial instinct,

some codified, memorial nucleic acid. They seek where they will find. A snake in the hand is not an object of repulsion, but a prize, a treasure. Crowds stare. The boy, the girl, the person who has found and caught the snake is focal: human with snake in hand, searched-out, overcome, mine now. Look. I did it. Could you?

Lazell paused, then recalled how he and Alexander followed paths that diverged and rewove, leading finally to Harvard. But in all those years Alexander had never heard Lazell tell the story of the King Snake—which, for Lazell, also began at the University of the South in 1957.

Lazell finished off a greasy chunk of coby, chipped a piece of ice off the block in the cooler buried in the sand, floated it, and began the tale of the elusive quarry that had drawn him repeatedly to the Outer Banks, like a moth to a lamp.

"Dr. Harry Yeatman, a professor of biology at Sewanee, had first told me about the big King Snake. When I entered the University of the South, I was just back from my first trip to the West Indies. I was full of the magic of islands and notions of discovery of new species, and rediscovery of species found long ago but dismissed as 'extinct,' or simply forgotten.

" 'I've got just the species for you,' Dr. Yeatman said. 'Someday, sometime, when you have the chance, look up the Ocracoke kingsnake. He was a real King Snake.' And indeed he was. I saw him a couple of years later—coiled, pickled in alcohol, in a big jar, at the Museum of Comparative Zoology at Harvard University. He is a massive, thick, dark brown, kingsnake banded with ivory white. He is heavily splotched and speckled on the head and anterior body. He has a broad-jowled, sharp-nosed head like a pine snake, not like a regular kingsnake. He is right there in his big jar, at Harvard, today. Harry Yeatman had known him alive.

"Yeatman was a student at the University of North Carolina at Chapel Hill when Dr. William Engels caught the Ocracoke kingsnake at the Knoll. He brought the big fellow back alive. Student Yeatman was his keeper. This serpent did not favor snakes, the customary fare of kingsnakes, but preferred rats. He slammed them up against the wall, like a pine snake, and did not constrict them, as ordinary kingsnakes do. Yeatman was as impressed by the behavioral differences of Engels's Ocracoke kingsnake as by its peculiar pattern and anatomy.

"I finished up my doctoral dissertation on iguanid lizards in the spring of 1968, seven years after I left Sewanee and Dr. Yeatman. But I had not forgotten the Ocracoke kingsnake. I knew him well. I went south and teamed up with my old friend, Dr. John A. (Jack) Musick, at the Virginia Institute of Marine Science. We headed south again, beyond Currituck Sound, to the Outer Banks.

"We took the ferry from Hatteras to Ocracoke, just as we do today. We consulted local folks and determined just where the Knoll was. We spent nine days hunting. Snake hunting is neither art nor science. It is a special alchemy made up of mud and sweat mixed with pieces of junk, rotted logs, back roads, and educated guesses. In hunting snakes, you turn over and look underneath. You turn over and look underneath everything you can find.

"On the Outer Banks, there isn't much junk to turn over, and there's only one road of any consequence to drive looking for DORs—animals Dead On the Road. Many of the type specimens of new species are first discovered Dead On the Road. Other people's road kills are the scientists' treasures. But during these nine days, we found few snakes of any sort and nothing like a kingsnake. In my field notes I wrote: *'If* sticticeps *[the Ocracoke kingsnake] lives, it's in rodent burrows, like a* Pituophis *[a pine snake].'*

"What rodent burrows? None had been found, none was known.

"That July we fixed a pattern that endured for many years. Jack and I would head out to the Banks, usually recruiting as many helpers as possible, and hunt snakes until we ran out of supplies. Back in those days you could not buy even a glass of beer south of Nags Head. We became canny campers. We learned the terrain and the good snake-catching spots. We hunted the high ground, looking for junk to turn over. That worked: we got racers, green snakes, ratsnakes, hognose snakes, glass lizards (which look like snakes), and assorted other species. We never saw a kingsnake.

"Finally, tired and frustrated, we would quit the Banks, cross the sound, camp at lovely Pettigrew State Park on the mainland, and engage in an orgy of snake-catching. That worked, too: in addition to dozens of others, we got magnificent kingsnakes. But, of course, those kingsnakes were just the well-known, garden variety, standard mainland sort of kingsnake, found from New Jersey to Florida. They looked nothing like Engels's Ocracoke King. The mystery deepened.

"We made five more trips like that one—in June 1969; May, August, and October 1970; and May 1971. Jack, a professional ichthyologist, had a regular job at the Institute back in Virginia; he could not spend weeks out here hunting. Alexander, a professional journalist, had a regular job in Piedmont North Carolina; he could snatch only a bit of vacation time. I, a secondary school biology teacher with a Ph.D., could not only take my school vacations, I could also organize official biology trips.

"I could load students into station wagons and my big

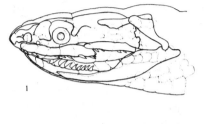

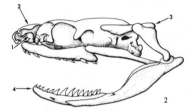

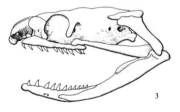

Kingsnake skull structure. 1) *L. g. sticticeps* skull within the head. 2) The skull of *sticticeps*: a) the inflated bulbous vomer; b) the heavy, thickened nasal bones; c) the massive tabular bone; d) the dentary bone. 3) A mainland kingsnake skull.

Chevy K-10 four-by-four and come south, all the way from Massachusetts. Jack and I became official National Park Service collaborators. We conferred regularly with the late Park Service naturalist Clay Gifford, a mine of information and encouragement. And I could easily and regularly see an Ocracoke kingsnake, too: every time I returned to Massachusetts there he was, coiled in his big jar, down in the basement of the Museum of Comparative Zoology. He is all there is.

"Harvard's late, great Thomas Barbour had teamed with William Engels to describe and scientifically name that lonely, strange snake. They called it *sticticeps*, which means spot-head, and calls to mind both the unusually heavy spotting and the peculiar head shape. We found regular kingsnakes in Buxton Woods and points north along the Banks. Some had heavy spotting, like *sticticeps*. None had that head shape. Most authorities suspected the lone *sticticeps* was merely a freak. I demand proof. Show me an ordinary kingsnake from Ocracoke, or a Buxton-type intermediate. But unless and until I have some Ocracoke specimen like that I will not make the terminal decision that *sticticeps* is invalid, a taxonomical nonentity. Not yet.

"I will not give up. I have cozened you to my camp. You have become hunters of King *sticticeps*. That is why, here in Hell, we are."

BUT THIS QUEST, this near-obsession, was no single-minded search for a single species. By 1971, when we began our mid-June expedition, we had come to realize that we were documenting at least two remarkably distinct ecosystems and finding other unknown animals. As we tromped the hammocks of Ocracoke we began to be profoundly impressed by how different their vegetation was from Buxton Woods or any highland north of Cape Hatteras. The dune vegetation was only a bit more subtly different. Even the marshes had a distinctive character.

Of course the land lies differently. South of Cape Point it runs southwest; north of Cape Point it comes down due south, or even southeast. Apart from artificial constructions, the dunes below Cape Point form irregular lumps and hillocks. To the north they form parallel ridges, often running essentially east-west. We had heard of a botanist, Paul Godfrey, working down on Core Banks, who had revolutionary new ideas. We had literature on plants and animals, and the encyclopedic mind of Clay Gifford to tap.

One of the worst problems was rodents. We were not plagued by them. Quite the contrary. We could not find them. South of Buxton Woods we could not uncover any small, native rodents that might supply a kingsnake with food. From Buxton Woods northward there were native mice and moles (moles are not rodents, but they might do for snake food). On Ocracoke there were only European introduced rats and mice (genera *Rattus* and *Mus*).

We dwelled on Dr. Yeatman's story of Dr. Engels's King Snake. Where had it learned to eat rodents? On Ocracoke? What rodents? The ones we humans brought it sometime after 1585? That notion would argue for a most remarkable sort of evolutionary history. Anyway, Ocracoke's introduced European rodents lived in the village, with their human victims. We hardly ever found evidence of them out in the bush. We certainly never saw or heard of a kingsnake in town.

And we had the Great Water Snake Problem. Water snakes we could find and catch. Not many, never enough, but we were finding some. Their taxonomy and evolutionary status were a mare's nest. We were determined to untangle it.

Camping on Ocracoke in mid-June 1971, we did well over the next few days. The girl, Numi, caught a magnificent four-foot water snake. The rest of us bagged racers, greens, hognose snakes, turtles, and lizards. But when the Park Service hauled us over to Portsmouth Island on June 19, 1971, the spell was finally broken, the mystery partially solved.

We were fanned out, working the marsh edge, hunting water snakes. Paul was point man. The point was a finger of wax myrtle and waterbush terminating in a bulk of storm wrack that included a wadded canvas tarp or sail. Paul began pulling it up. He was fifty yards away.

He came up shouting. He came up waving both arms above his head, spread wide, hands clutching and waving more than five feet of mahogany-and-ivory-banded snake. Big snake: King Snake.

It seemed a transposition. It was as though the old jar in the basement of the museum in Cambridge had been uncorked by remote control and Engels's serpent beamed into the hands of the hunter. After more than thirty years, here was that old, dead snake's twin. Here was *sticticeps*, alive.

We thought it a freak. It was in the wrong place, of course. We knew that because we knew lots about kingsnakes. Although they sometimes frequent the edges of freshwater marsh, they typically prefer slightly higher, drier terrain. They are rarely, if ever, found along the edges of saltwater marsh.

Three days later, back on Hatteras Island, we picked up a road-killed *sticticeps* beside a black rush marsh just southwest of Billy Mitchell Airport. Two days after that we found another dead specimen, the fourth known *sticticeps*, on a little road going through a black rush marsh in the town of Hatteras.

We performed a mental transformation. We conceded that we had spent three years looking in the wrong places. We quickly bade goodbye to conventional, textbook notions; this was a snake of the salt marsh.

But this discovery kindled a bigger mystery, transported us—as they say—back to basics. What did *sticticeps* eat out there in the salt marsh? Remember Harry Yeatman's information: this was a rat-eating sort of snake.

In October we found two more *sticticeps* under wrack in the salt marsh. We brooded over the mystery through the gray, dark winter, awaiting the spring of snake-hunting. Facts, for scientists, are light in the darkness. The more facts, the larger the area of light. The larger the area of light, the greater the edge of darkness upon which it touches. But we had inspired an illuminator. The girl, Numi, tromped the edge of darkness in her mind. The problem obviously revolved around our lack of knowledge about *mammals*— furry, warm creatures that a cold and glistening *sticticeps* kingsnake would like to eat. There were lists and literature records of mammals for the Outer Banks, but they were not very enlightening. Apart from European rats, largely confined to human edifices, there were sparse and scanty records of cotton rats, cotton mice, harvest mice, voles, moles, and rice rats.

Most were apparently very rare or very local. Cotton mice (or, as it has since turned out, their close relatives, the white-footed mice) were fairly common in Buxton Woods, just north of the great elbow of Cape Hatteras. So were moles— there and on northward up the Banks. Neither occurred *south* of the elbow, in the area where we found *sticticeps*.

Kingsnakes roamed in Buxton Woods, and northward too, but they were not the *sticticeps* variety. Some bore a resemblance to *sticticeps*, but none was the real snake itself. No small, native rodent was known in any numbers from the range of *sticticeps*. The only plausible, even possibly suitable, rodent to fill the needs of *sticticeps* might be the "marsh rice rat," or *Oryzomys palustris*, but no solid evidence existed.

Numi had been raised by snake hunters. She naturally assumed that the way to catch things—anything (well, maybe not elephants)—was to look under debris. You find a thing like a board, you turn it over, and you find a thing like a snake. We did spot rodents this way, too, but we did not catch them. We were not interested in them. Anyway, they moved too fast. If you did catch one it would surely bite. Silly as it sounds, snake hunters worry much more about rat bites than snake bites.

So, in a crowd of compulsive snake hunters, Numi set out to become a mammalogist. On May 15, 1972, camping at Cape Point, Hatteras Island, she wrote in her field notes: "Today we are leaving for Ocracoke Island. I am probably going to camp there because catching a cricetid [ratlike] rodent there is more important than catching a mole here."

It rained. The wind came up. A real spring northeaster. We retreated to the mainland. Numi's notes of May 1972: "No luck but I will take care of that this summer."

An aside: On the mainland two miles west of Grapevine Landing in the Alligator River swamps, Numi recorded on May 19, 1972: ". . . right front paw print of *Ursus americanus* [a bear] for sure. According to B&G [a popular field guide to mammals] it is out of its range." If Numi could find bears overlooked by the writers of mammal guides, she ought to be able to find rats, too, we reasoned.

On July 7, 1972, turning sheet metal in the salt marsh back on Ocracoke, she did: "*Oryzomys palustris* . . . Caught by hand. Yeah!" She had the rice rat from the range of *sticticeps*. More light in the darkness. Numi bought some rat traps.

Even though rice is a marsh grass native to Asia, rice rats, she knew, are a strictly American group of small mammals. The species name translates directly from Latin: *Oryzomys* means rice mouse and *palustris* means "of the marsh." These rodents are larger than the native white-footed mouse, and not so boldly patterned or richly colored. They are similarly larger than the somber European house mouse, but smaller than the noxious pest European rats that have been introduced throughout North America by humans. They are brown above, with slate to near-black guard hairs, and paler below. They have valved nostrils and are agile swimmers, like their much larger relatives, the muskrats.

For several years afterward Numi still hunted mammals like a snake hunter: turn things over and grab. Like a seasoned, trained, stoic snake hunter, she did not let go of things that bit her. *Hold it! Do not let it go! Once it bites you it's done its worst—and best. If you let go it wins, you lose!* Numi would catch mammals, turning junk by hand, until she found something interesting. Then we would pile into the truck and drive to the nearest store that sold rat traps and buy some. Snake hunters are not renowned for planning ahead.

We moved to Portsmouth. Numi began to tear up wrack mats along the salt marsh edges. She uncovered miles of rice rat runways. She caught a plentiful supply of rice rats. She found rice rat tunnels that led out of the marsh to very large, old wax myrtle trees. Under the root platforms of the wax myrtles it appeared that the rice rats had excavated chambers. She worked hard for two whole days before she found her first kingsnake.

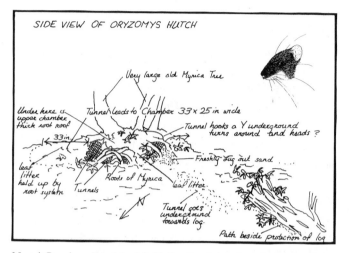

SIDE VIEW OF ORYZOMYS HUTCH

Very large old Myrica Tree

Under here is upper chamber thick root roof

Tunnel leads to Chamber 33 x 25 in wide

Tunnel hooks a Y underground turns around and heads ?

33 in

Freshly dug out sand

leaf litter held up by root system

Roots of Myrica

Tunnels

leaf litter

Tunnel goes underground towards log.

Path beside protection of log

Numi Goodyear's original field sketch of chambered wax myrtle (*Myrica*) made at Portsmouth, July 1972. The rice rat, top right, illustrated her first published paper in March 1973.

On July 9, 1972, Numi found a big *sticticeps* kingsnake in the rice rat runs within a few feet of a chambered wax myrtle. We all began to look. By July 18 we had found two more. It worked: *sticticeps* live in rice rat tunnels; rice rats live in the salt marsh. But what about those big wax myrtles? Why were they dug out to form chambers underneath?

Recall Dr. Paul Godfrey's study of oceanic overwash, and how the Outer Banks are built and rebuilt. If the Banks depend on overwash, the animals must survive it. The abundant ones and the endemic ones—those found in big populations and those found nowhere else, respectively— must survive it very well, virtually depend on it themselves.

Good science starts in the field, of course, where specimens are gathered and facts discerned. But there is more to the process than that. One must try to make sense, scientifically speaking, of what one has found, or not found. It is in the evenings, when one is sequestered from the biting flies and mosquitos, enjoying shrimp steaming on the stove and oysters frying in the clear, rendered, ham-fat oil, that such reflection often takes place. A lot of hard science gets done during such nocturnal reckonings of the day's—or the week's, or the lifetime's—observations and collections.

Science begins with those collections and observations— facts: light illuminating the widening edge of darkness. Next comes hypothesis, theory: *What if? What if the wax myrtle root platforms are airtight above?* What if they trap air during overwashes? What if the rice rats retreat to them when all the rest of their world goes underwater? That is mostly during late summer (hurricane) through winter (northeaster), and rice rats are scarcer and harder to find in the spring than in summer.

What if the only rice rats that survive overwash are those close enough to a chambered wax myrtle to seek refuge there?

If that were true, it would logically follow that the endemic kingsnake—found nowhere else in the world, totally dependent upon rice rats—could not survive at any great remove from a chambered wax myrtle either. Young kingsnakes might disperse in both directions along the wrack lines and marsh edge, but long-term survival—survival to reproductive adulthood—would depend on a good supply of food. Food equals rice rats—rice rats in spring and early summer when energy demands are high because winter draws down the fat-energy supply and reproduction needs energy.

What if a kingsnake in the Intercapes Zone of the Outer Banks of North Carolina needs spring rice rats? It had better live in close proximity to a chambered wax myrtle tree.

≈

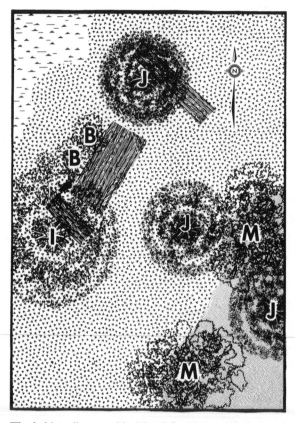

The habitat discovered by Numi Goodyear at Portsmouth in the fateful summer of 1972. Top left is the rush marsh. Lower right is high ground. *B* is waterbush, genus *Baccharis*. *I* is yaupon, genus *Ilex*. *J* is redcedar, genus *Juniperus*. And *M* is wax myrtle, genus *Myrica*. Note boards in wrack debris.

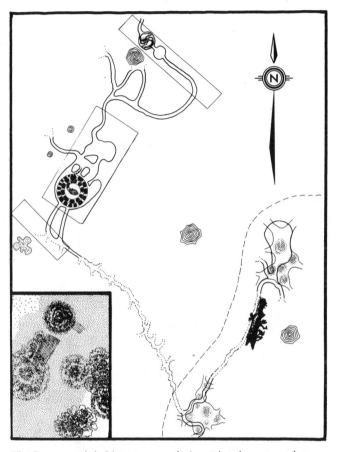

The Portsmouth habitat uncovered. An active rice rat nest is top center. Well-formed runways and chambers are indicated by bold, solid lines. Vague or diffuse runways are dotted. While both large wax myrtles show runways, only the biggest, right center, was chambered. Inset shows original habitat. We replaced the nest as closely as possible after Numi Goodyear caught the kingsnake, left center.

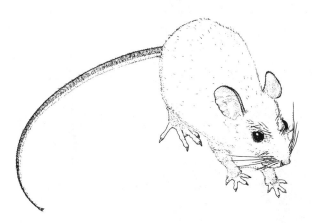

Marsh Rice Rat

THUS HAD NUMI conceived a real scientific hypothesis over fresh shrimp and ham-fried oysters. Next, in science, you must test your new hypothesis.

On July 22, 1972, Numi set out to test her hypothesis by reversing the search pattern. She scanned the marsh edge for big wax myrtles. When she spotted one she crashed her way to its base and checked for tunnel entrances into a chamber. When she found the chamber she set out along the nearest wrack line in the black rush looking for a kingsnake. She found one in four hours.

We headed down Core Banks, into territory we—the snake hunters—predicted, based on our hypothesis of kingsnake range in the Intercapes Ecological Zone, should bear *sticticeps* kingsnakes. Numi rode on the roof of the truck. She scanned the marsh edge for big wax myrtles. She pounded on the roof when she spied one she liked. She brush-busted over to it and whooped and waved and yelled if it was chambered.

We all piled out of the truck and ran for the wrack line, the marsh edge. Sometimes we got there before Numi found the kingsnake, so sometimes one of us would find one. Twice we found two. . . .

We set about compiling more facts. We went to the edge of Blackbeard's Hammock at Springer's Point on Ocracoke. The very same processes of shoal development and channel shift that nearly killed Lieutenant Maynard in 1718 were still at work. The growing shoal had shifted the channel southward, eating into the edge of the land, washing out trees. We could pick and choose among dead specimens.

We examined washed-out cedars, yaupons, waterbushes, and even live oaks. We studied them by sawing off their branches and roots, hauling them up on land, and turning them upside down. Then we poured water into any cavities they had that might have held air when right-side-up. They could not hold anything. We never found any sort of tree or bush in the Intercapes that could be made into an airtight chamber that was not a wax myrtle.

Most any wax myrtle seems to form an umbrella or parachutelike root mass. Lots leak because the amalgamated root stocks do not web together completely. Older, bigger ones are the most likely candidates. The biggest, best, tightest one we could find among those washed out of the edge of Blackbeard's Hammock held two liters without leaking. It was about half the size of those we found chambered in the field.

By probing with a wire, after removing the soil, humus, and leaf litter, we could at least determine the expanse of solid wood above a wax myrtle chamber *in situ*—alive and in the field. We could not determine how much air it traps without digging it up and killing it, thereby destroying that entire unit of the hypothesized ecological, evolutionary system.

There was another way to test the hypothesis, but it was much more taxing. That was to try to determine if the rice rats really went into the chambers during overwash. Now how would Numi possibly do that?

First, she would need to be ensconced on the Banks during a big storm. (It is notable how many scientists seem to want or need to be out there during a big storm.) Second, she would have to sense the presence of little, furry, warm mammals underground, through solid wood, *underwater*. But how are you going to *do* that?

Simply measuring water levels is tricky enough. Numi devised two sorts of tide or depth gauges. The first was a pole driven into the bottom at marsh edge with little cups attached every two centimeters. Rising water fills the cups. However, rain can, too, and the rising water in which we were most interested—storm flooding—was typically accompanied by rain.

The problem was not insurmountable: we could check the salinity of the water in the cups. Wave action could slop some seawater into cups higher than the standing water actually reached, but by observing this in rough water locations and getting experience day after day with regular high tides, we soon enough learned to pick up the salinity changes necessary to correct for this effect. Finally, the system was cumbersome because one had to empty all the cups after each high water to start over. This required pulling up the rig, dumping it, and resetting it. A level line, just like those used in construction, was set up from the tide gauge station to the base of the chambered wax myrtle. The gauge had to be reset to the exact same depth in the bottom relative to the level line each time, too.

Numi invented a much more elegant gauge using two pieces of clear plastic pipe sealed together, side by side, with aquarium cement. One pipe was kept dry by capping it at both ends. It had one small hole in the side near the top, through which passed a monofilament line into the second, wet pipe. The wet pipe was open at the bottom and top and riddled with holes. A cork float in the wet pipe was attached to a split-shot weight in the dry pipe. As the water rises in the wet pipe, the weight drops in the dry pipe. Wave action and rain are ineffective in deluding the observer. The whole contraption is reset simply by pulling the monofilament on the wet side back through the hole. It was a dandy device.

Next, we had to sense the presence of the rats in the chamber. Rats are mammals: they generate heat. So, Numi acquired some maximum-registering thermometers: they tell how warm it got even after the place has cooled down again. She would bore holes through the solid wood of the wax myrtle root masses, insert the thermometers through rubber stoppers (made to hold glass tubes), and drive the stoppers into the holes.

We arrived in Ocracoke on March 19, 1973. Between then and April 24, Numi monitored ground temperatures in several control chambers inland from the marshes, air temperatures, sound water temperatures, temperatures in six different wax myrtle chambers, and high water. Nothing much happened over most of that time. There were normal, regular high tides twice a day; temperatures in control chambers and in wax myrtle chambers stayed within a degree of each other. Only once did anything dramatic occur.

On March 27 a northeaster hit on the rising tide. The water rose eleven centimeters—over four inches—above normal high tide, flooding the marshes to above the level of the root platforms of the two wax myrtles Numi was then monitoring. The temperature inside the wax myrtle chambers shot up 2.5 degrees centigrade (about 5 degrees Fahren-

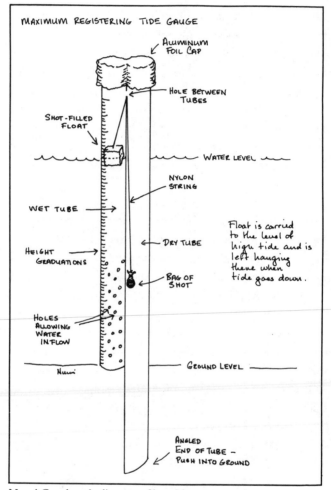

Numi Goodyear's diagram of her tide gauge.

heit) above the previous record high, and more—3.5 degrees centigrade—above the temperature in the control chamber at that time. Something warm scurried into those chambers during that flood.

ECOLOGISTS AND PHYSIOLOGISTS have recently focused a great deal of attention on the phenomenon of *coevolution*. Actually, *phenomena*, for no one example of coevolution is quite like another. The textbook definition of coevolution is this: *In each case where it occurs, two or more kinds of contemporaneous organisms evolve adaptations resulting in interactions so close that each exerts a strong selective force on the other*. Each phenomenon depends critically on the other: contemporaneous organisms, adaptations, interactions, strong selective force. Without any one part of the equation, coevolution does not occur.

As always in science, specific examples illuminate. Obvious examples of coevolution include pollinators and flowers, where both a pollinator (usually an insect) and a specific flower have become so modified that each fits the other perfectly, usually to the exclusion of other species. Food plants and the herbivores that eat them may undergo coevolution, where a chemical produced by the plant is needed in the repertoire of one specific herbivore but is toxic to other species that might otherwise munch that plant. The monarch and the milkweed make a fine example of coevolution that also happens to be beautiful to the eye. The monarch—a big, showy butterfly—eats milkweed as a caterpillar, a plant that produces a viscous white sap highly toxic to almost everything else. Milkweed sap is mother's milk to the monarch, however, for it is immune not only to the toxins, but also actually incorporates them into its own chemical makeup to poison would-be predators.

Enter a third species, the viceroy—a smaller butterfly that looks like a monarch (at first glance). The viceroy cannot eat milkweed, but because some butterfly predators mistake it for the monarch and "know" monarchs are toxic, many viceroys escape predation: classic mimicry.

To claim this is a three-species coevolutionary system, we would need to close the circle and argue that some benefit derives to the milkweed from viceroy biology. Coevolution is, after all, a two-way (or circular) street. Every partner in the coevolutionary pact depends critically on the other. Well, selecting for butterflies that look like milkweed eaters but do not actually eat milkweed will tend to benefit those mimics and in turn fill the herbivore niche with individuals that are harmless to milkweed. The fetch of that argument might seem a bit far, but *any* selective benefit, no matter how slight, will prevail ultimately in a species of reasonably fecund survivors. It is almost certainly to the milkweed's benefit to increase viceroys, because the viceroy caterpillar does not dine on its leaves.

It may be useful at this point to review just how Darwinian evolution works. It works by natural selection, but that is not different in effect from the artificial selection humans perform to make breeds of dogs or cultivars of useful plants. The species undergoing selection—*all* successful species— produce far more offspring than are needed to maintain a population. Most of these offspring are culled out, never reach reproductive age, or simply fail to reproduce. The vicissitudes of life eliminate any chance that a well-adapted, stable population can exist by simply producing offspring at replacement rate. There *must* be "waste" of offspring, and that "waste" *must* be profligate. (Actually, of course, there is no "waste" at all ecologically: all life depends on the harvest of someone else's surplus.)

In all known universal environments at least two processes militate against "running in place" evolutionarily without profligate "waste" of offspring. First, the environment changes. No matter how well you and your mate are adapted to today's environment, tomorrow's will select for something different. Next, there is the Second Law: entropy. The Second Law of Thermodynamics states simply that entropy is inexorable: everything breaks down. A perfect genetic code, even in an unvarying environment, will not be perfectly replicated very many times. To get any perfect—or even nearly perfect—copies you will have to make a good batch of extras. Then expect lots of them to die off—unless environmental change enables one of those "imperfect" copies luckily to fit a new niche. (In which case you, the original progenitors, are now out of luck; the world has passed you by.)

The mechanics of coevolution are the same as those for any other examples of natural selection, except that two or more species are seen to be evolving in concert. In the putative case of the Intercapes trilogy—kingsnake, rice rat, and wax myrtle—natural selection must have been remarkably rapid, since the Outer Banks as we know them have existed for only a few thousand years. Charles Darwin noted: "The periods during which species have undergone modification . . . have probably been short in comparison with the periods during which they retained the same form." Rapid bursts of evolutionary change set in a long time matrix of relative stasis are common and typical of the process. The process has today been given the catchy name "punctuated equilibrium," but it remains exactly as Darwin described it: not a new idea.

But the Intercapes Zone is a geologically new theater for evolution. It is distinctly different from the Carolina mainland, of course, but also from other portions of the Outer

Banks. To perform its role, the kingsnake is the species that seemingly underwent the most radical, adaptive, evolutionary shift—and very fast, while its relatives elsewhere did nothing of the kind. Here we see a punctuation *within* the equilibrium of a widespread species (ranging from New Jersey to Florida, Mexico, and California), involving only those populations in one small geographic segment of the range. Natural selection must have been brutal to forge the *sticticeps* form, and must remain brutal to maintain it. A spectrum of offspring may be produced, but only a tightly circumscribed portion of those—just right for the Intercapes environment—succeed, survive, and carry on the breed.

It seems obvious that the kingsnake is specifically adapted to a primary prey species, the rice rat. And with equal clarity we see the adaptive dependence of the rice rat on the wax myrtle. Thus, indirectly, the kingsnake is adapted to the wax myrtle, too. We suspect and suggest that a direct adaption exists as well: that wax myrtle chambers serve as kingsnake hibernacula—winter refuges. To close the loop we must perceive some benefit of these animal dependencies deriving to the wax myrtle that so kindly provides watertight chamber roofs.

Many myricaceous plants, like the more familiar leguminous plants such as peas, clover, and soybeans, fix nitrogen. *Myrica cerifera*, the wax myrtle, is no exception. We all need nitrogen to make amino acids and, in turn, proteins. There is plenty of nitrogen in the air, but few ways to "fix" it in chemical bonds so that we can use it. A few plants—like locust trees and wax myrtles—and even fewer animals—like termites, cows, and kangaroos—harbor microorganisms such as bacteria that fix nitrogen. All the rest of us must eat it in the form of proteins or—in the case of plants—draw it up into tissues in usable form, already fixed for us by some other organism.

The obvious benefit for a plant in providing a refuge for vertebrate animals would be fertilizer from the animal's urea: a rich source of nitrogen. However, nitrogen from rat and snake urea would be little benefit to a wax myrtle, harboring as it does its own nitrogen fixers. At first we despaired of closing the circle. We could show a chain of dependence: kingsnake on rice rat on wax myrtle, but that was it. Then we consulted some plant physiologists, and the circle began to close.

Wax myrtles grow best right at the interface of the fresh-water lens and the salt marsh. The only other shrub that flourishes there is waterbush (*Baccharis halmifolia*), a composite or daisy family plant despite its woody trunks. But waterbush is a profligate reproducer of short life; its trace mineral needs are not so long or resource-exhausting as those of a long-lived wax myrtle. The interface where wax myrtles live is one dominated by outward seepage of nutrient-poor water from nutrient-poor sand. The opportunity to obtain magnesium, potassium, phosphorus, and cobalt, all accessible in seawater, could give a real selective advantage to individual wax myrtles able to—literally—grasp it.

Marine invertebrates such as the small snails and arthropods of the salt marsh often contain ample amounts of trace minerals. They are favored foods of the partially carnivorous rice rat. Rice rat feces directly deposited around wax myrtle roots—especially where a chamber enables some seawater to pool during overwash—might convey significant advantage to chambered individuals.

Certainly wax myrtles grow bigger and live longer than other plants in their low-nutrient habitat. All the woody species there, or close to there, produce a shallow, splayed root mass that taps the widest possible nutrient bed and gives maximum upright stability. These species include waterbush, yaupon, redcedar, and even live oak. No other provides rice rat chambers. No other thrives as well at the nutrient-poor interface. The biggest—so we suppose oldest and most successful—wax myrtles are chambered, although some small ones with good root platforms are, too.

We present a testable hypothesis: rice rat feces provide a direct bridge for trace elements from the sea to wax myrtle root masses; those wax myrtles best configured to offer airtight chambers for the rice rats benefit most. Any hereditary tendency (gene) wax myrtles have to produce the best root system for chambering will be selected for.

It is a near certainty that rice rat chambers do not harm the wax myrtles, or the biggest and best would not have them. If our hypothesis proves correct, then the Intercapes trilogy would constitute one of the most remarkable examples of coevolution yet seen: two vertebrate species—a reptile and a mammal—and a plant, all coevolved in an ecosystem less than 5,000 years old. That is rapid even by modern Darwinian standards, and it confirms the importance of the Outer Banks as a laboratory for further scientific exploration.

OVER A YEAR LATER, in 1974, Numi Spitzer graduated from high school. It would be more than fifteen years before we revisited the curious questions posed by wax myrtles, rice rats, and *sticticeps* kingsnakes in the Intercapes Zone of the Outer Banks. By then she was not simply a rice rat catcher among snake hunters; she was Dr. N. C. Goodyear, a professional biologist. Another circle closed, or nearly so.

Chapter Seven
Flight

. . . I would be particularly thankful for advice
as to a suitable locality where I could depend
on winds of about fifteen miles per hour
without rain or too inclement weather.
I am certain that such localities are rare.
Wilbur Wright, 1900

Neither you nor I
Ever thought to fly.
Oh, but fly we did,
Literally fly.
Robert Frost, 1953

A N ENDURING, ENDEARING improbability envelopes the legend of the brothers Wright: that two eccentric, self-educated bicycle mechanics from Dayton, Ohio, could conceive, build, and successfully launch the world's first powered, sustained, heavier-than-air flying machine. Fortunately, serious studies of their historic achievement in recent years have debunked some of that legend. This debunking has made the Wrights' success even more extraordinary. It has been shown to be the result not of amateur tinkering, but of tedious work, perseverance, keen intelligence, courage, and teamwork—all resting solidly on the shoulders of aeronautical research that had preceded them. There were no amateurs here, no blind luck. The Wright brothers knew exactly what they were doing, and what success would mean for their place in history.

It may seem slightly improbable, too, that these straight-laced sons of the Midwest would be drawn to Kitty Hawk, North Carolina—a remote, desolate, sand-swept outpost reachable only by boat. Odd that they would not only come here, but also that they would survive and even flourish in such an alien environment. Yet on closer examination, it was a good fit—like the well-tailored clothes the sartorially conscious brothers wore.

As with the adventurers who preceded the Wrights and the scientists who followed them, the Outer Banks served as a living laboratory in which to test their ideas and to realize their dreams. They took from the Banks what they needed; the Banks gained from them a firm niche in history. The Banks now qualify as one of those geographical frontiers where landscape and scientist or engineer are fused in the imagination, like Darwin's Galapagos Islands or the astronauts' Cape Canaveral. Similarly, the Wright brothers' Kitty Hawk will be forever remembered as the birthplace of flight, of man's liberation from purely earthly pursuits.

There were other locales, no doubt, where Wilbur and Orville could have conducted their glider-flying experiments. The Outer Banks own no patent on the wind. But the unusual confluence of wind, sand, and open spaces made the Outer Banks ideal for the Wrights' purposes. As William Tate, postmaster at Kitty Hawk, wrote the brothers in response to their inquiry about conditions there: "In answering I would say that you would find here nearly any type of ground you could wish. You could for instance get a stretch of sandy land 1 mile by five with a bare hill in center 80 feet high not a tree or bush any where to break the evenness of the wind current." Apart from its physical compatibility with the brothers' criteria, the Banks' stark setting seemed a logical extension of the Wright brothers' personalities: self-contained, stoic, isolated.

When Wilbur and Orville first came to Kitty Hawk in September 1900—Wilbur was thirty-three, Orville, twenty-nine—they had no notion that their experiments would prove successful. They were initially drawn, of course, by reports from the U.S. Weather Bureau that Kitty Hawk claimed the sixth-highest average wind in the United States, at 13.4 mph. Moreover, Kitty Hawk offered a relatively soft padding of sand from which to launch and land their glider, and it was far removed from the curious and unwanted spectators of a typical urban area. But Kitty Hawk also fulfilled the Wrights' desire to visit an exotic, faraway place.

"It is my belief that flight is possible," Wilbur wrote his father in early September 1900, "and while I am taking up the investigation for pleasure rather than profit, I think there is slight possibility of achieving fame and fortune from it. . . . At any rate I shall have an outing of several weeks and see a part of the world I have never before visited." It was the brothers' first trip away from home in seven years. Wilbur later told his father that he saw his time at Kitty Hawk as "a pleasure trip, pure and simple, and I know of no trip from which I could expect greater pleasure at the same cost."

This spirit of adventure continued to animate the Wright brothers' yearly sojourns to the Outer Banks. Where less-resilient souls would have quickly abandoned their work in Kitty Hawk's often uncharitable environment, Wilbur and Orville seemed challenged and invigorated by it. What better metaphor for their undertaking than those bleak photographs showing their tent lashed to a scraggly oak tree and surrounded by sand as far as the eye could see?

For them such isolation was blissful; with a few exceptions, they found visitors a source of irritation and hindrance. The interruptions simply slowed their work. Wilbur and Orville preferred the company of the hearty Bankers who helped them carry their gliders up the dune and back again, day after day. Few of the personalities who have shaped the history of the Outer Banks better suit the islands' ecology than the Wright brothers; none has taken more effective advantage of it. What for most would have been an empty, sterile landscape became for them a runway to the sky.

BEFORE HARNESSING THE wind in their ambitious pursuit, the Wright brothers first came to respect its properties and its power. As we shall see, the Outer Banks proved to be a relentless and sometimes brutal tutor on the subject of wind. Unlike most other aviators of their day, the Wrights did not try to tame the wind, nor to overwhelm it. Their strategy was sound: first, use unpowered gliders to test the behavior of man-made wings in winds of varying direction, steadiness, and velocity. Once these secrets of control had been unraveled, an engine could be attached to drive the machine through air both thick and thin.

All life on earth depends on wind: the movement of our atmosphere, uniquely rich in free oxygen and convertible carbon and nitrogen. Yet wind is not a thing in itself. It is a process. It is a relationship between things moving at different speeds, between gases and vapors, vapors and liquids, gases and solids.

Without the process of wind there would be no process called weather. There would be heat and cold, but they would be unmitigated, immutable. Apart from the occasional tectonic event—a volcano or an earthquake or a collision—nothing would happen. There would be no Outer Banks. Nothing on this planet would ever fly. With the wind, however, all things are possible, including flight. Wind is an integral part of the dynamic, organic concept of the earth known as Gaia. In the words of Lyall Watson, author of *Heaven's Breath*, "This concept of an organic universe, of a cosmos that lives and breathes, is an important one. It carries the seeds of a new and more tuneful understanding. One rich with the eloquent cadence of the wind."

In simplest terms, wind is generated by the sun's rays: heat absorbed by earth that is immediately conducted to the air molecules on the surface. Heat agitates the air molecules and makes them move. Where earth gets hottest, at the equator, enormous amounts of hot air rise. Rising ever higher, the air masses cool. Once sufficiently cooled, the now-slowed, now-calmed molecules condense. The air masses, denser now, become relatively heavy and so descend.

Where the air comes down, about 30° N in the Northern Hemisphere, there is as little lateral air movement as where the air rises at the equator. These seemingly windless regions are called the Horse Latitudes at sea. In the days when sailing ships transported horses, it was here one often had to jettison the cargo to lighten ship. Lighter, the ships could be towed by men at oars in small boats until a breeze moved again. On land, the major deserts of the world are centered on these latitudes. Here, fronts between laterally moving air masses of distinct temperatures and pressures rarely stray. The weather is largely distinguished by its absence.

The cooled air that reaches the surface at the Horse Latitudes pushes both north and south. The portion moving northward provides the southern component to the Outer Banks' prevailing southwesterlies. Their westerly component is the result of Coriolis Effect. It is named for Gaspard Coriolis, a French engineer of the 1840s. Perhaps Gaspard discovered this Effect one Sunday morning while riding a merry-go-round horse. Perhaps he tried to throw a ball to a friend, who was riding another horse on the opposite side of the merry-go-round. Anyone can discover the Effect that way.

What the Coriolis Effect does to our northward traveling air in the middle, temperate latitudes—where the Banks are— is deflect it east, which makes a westerly wind. What actually occurs is that the speed of the earth turning under the north-moving air decreases because the earth gets smaller in diameter as one travels away from the equator. By doing nothing but travel north, the air, relatively speaking, gains speed from the west. At the equator the earth is spinning a thousand miles an hour. At the north pole it is going zero. At Cape Hatteras it is going about 380 mph. This translates into a major relative speed gain from west to east for the north-moving air.

To this relatively simple picture vast complexities are

added. Cells of very hot air drift westward off of equatorial Africa. Already these cells contain thin air—air of little density because it is so hot. Air in these cells is already rising. When a cell moves over warm water, however, it takes on water vapor. In gaseous form, water molecules are very light: one big oxygen atom attached to two tiny hydrogens.

The light water molecules displace the heavier molecules of nitrogen (two big, heavy nitrogen atoms) and oxygen (two big, heavy oxygen atoms) that make up 99 percent of air. The result is an even lighter, less dense, mix, ever so much more upwardly mobile.

As a hot, wet, light cell drifts north, the Coriolis Effect puts a spin on it. That is how hurricanes are born. But hurricanes are sometime things; they do not linger. The Gulf of Mexico is another warm water mass. It lies precisely where the Banks' prevailing southwesterlies are apt to begin. It, too, can send up a burst of warm, wet, light, air, the edge of which is a low pressure front. And, away up north over Canada, the air gets very cold. Cold air is dense and therefore heavy. Cold, dense air is apt to flow outward from Canada, sometimes pushing away the southwesterlies with a blue norther, the Montreal Express. The forward edge of cold, dense air is a high pressure front.

The Outer Banks are ideally situated to experience all the caprice that winds and weather can muster. As denizens of the middle latitudes, they bear the brunt of frequent pressure systems clashing together. As inhabitants of the open sea, they are exposed to the harshest weather without the moderating effect of a mountain range or even a respectable forest to slow it down. And at Cape Hatteras, the collision of two major ocean currents only adds to the general turbulence. Pressure systems often stall over the Banks, held there by the Gulf Stream's invisible hands.

No wonder, then, that fickle winds are so much a part of life and lore on the Outer Banks. As author and Outer Banks resident Jan DeBlieu has written, "Wind is culture and heritage on the Outer Banks; wind shapes earth, plant, animal, human. It toughens us. It moves mountains of sand as we watch. It makes it difficult to sleepwalk through life."

WILBUR WRIGHT DID not take long to get his proper introduction to Kitty Hawk in that September of 1900. It arrived in inimitable style: a fierce storm, perhaps the remnants of a Gulf Coast hurricane, nearly swamped the schooner on which he crossed Albemarle Sound.

Passage to Kitty Hawk at the turn of the century was possible only by boat. On September 11, 1900, Wilbur departed from Elizabeth City, a small town on the Pasquotank River, in a skiff piloted by Israel Perry. Perry explained that he owned a flat-bottomed schooner anchored near the mouth of Albemarle Sound. The trip would ordinarily have taken five to six hours in fair weather—even in a vessel as leaky and decrepit as Perry's. As Wilbur described it, "The sails were rotten, the ropes badly worn and the rudder post half rotted off, and the cabin so vermin-infested that I kept out of it from first to last."

It was nearly dark when Perry, Wilbur, and a third passenger, a boy, finally set sail, in a light wind. But calm conditions did not prevail for long. "The water was much rougher than the light wind would have led us to expect," Wilbur wrote. "Israel spoke of it several times and seemed a little uneasy." And well he should. The wind shifted to the south and east, forcing them to face an increasingly strong headwind. Water pouring over the bow necessitated frequent bailing.

Just before midnight, as they were struggling to round the

north shore and seek shelter in the North River, a strong gust ripped the foresail from the boom. Wilbur wrote in his journal that he and the boy succeeded in taking it in, but not without peril as the boat rocked to and fro. Wilbur continued:

By the time we had reached a point even with the end of the point it became doubtful whether we would be able to round the light. . . . The suspense was ended by another roaring of the canvas as the mainsail also tore loose from the boom, and shook fiercely in the gale. The only chance was to make a straight run over the bar under nothing but a jib, so we took in the mainsail and let the boat swing round stern to the wind. This was a very dangerous maneuver in such a sea but was in some way accomplished without capsizing. The waves were very high on the bar and broke over the stern very badly.

At last Perry reached lee shelter in the river, and all three collapsed on the deck, exhausted and drenched. Perry's boat, appropriately, was named *Curlicue;* its gyrations Wilbur would not soon forget. They spent the rest of the night at anchor and made repairs until midafternoon the next day. The wind shifted and permitted them to continue their voyage, until they reached Kitty Hawk Bay at dusk. It had been a harrowing trip that would rival any Wilbur would subsequently make by air.

The village of Kitty Hawk into which Wilbur walked the next morning consisted of a handful of houses and stores clumped near the bay. Located on the beach across from the village were a Weather Bureau office and a two-story lifesaving station. It would be misleading to suggest, however, that these few buildings and their occupants comprised the whole of human habitation in the area. Just twelve miles to the south was Nags Head, a resort town which served as a popular summer retreat for well-to-do North Carolinians.

But separating the two communities was nothing but a narrow expanse of sand, punctuated by large sets of dunes. The nearest of these to Kitty Hawk were the three dunes of Kill Devil Hills, four miles to the south. The dunes, the highest on the Outer Banks, are maintained by storm winds from the northeast and, from the opposite direction, the prevailing southwesterlies. Over time they are pushed, or rolled, from sea to sound, like giant sentries in retreat.

Years later, Orville told an artist who asked him to describe the setting that it was "like the Sahara, or what I imagined the Sahara to be." The bleakness of the setting was not improved by the extensive nibbling and grazing of the Bankers' nomadic domestic animals, who prevented vegetation from assuming its rightful place among the dunes. As Orville wrote to his sister, "You never saw such poor pitiful-looking creatures as the horses, hogs and cows are down here. The only things that thrive and grow fat are the bedbugs, mosquitoes, and wood ticks."

In a letter to flying enthusiast Octave Chanute, the brothers' friend and chief cheerleader, Wilbur conveyed an accurate overview of the place:

We located on the bar which separates Albemarle Sound from the ocean. South of Kitty Hawk the bar is absolutely bare of vegetation and flat as a floor, from sound to ocean, for a distance of nearly five miles, except a sand hill one hundred and five feet high which rises almost in its center. The main slope of the hill is to the northeast, which is facing the prevailing winds. The slope is one in six. . . . To the north, northeast, east, and southeast there is nothing but flat plain and ocean for a thousand miles nearly. It is an ideal

place for gliding experiments except for its inaccessibility.
The person who goes there must take everything he will
possibly need, for he cannot depend on getting any needed
article from the outside world in less than three weeks.

To most outsiders, the dunes have represented, at the
least, structures to be gazed at in awe; at the most, play-
grounds for kite-flying and, more recently, hang gliding. But
for Wilbur and Orville Wright these dunes, their precise
slopes and their interaction with the wind, were of keen
interest: not only the brothers' reach for fame, but also their
lives, depended on the accuracy of their calculations.

The house toward which Wilbur walked that September
morning belonged to Bill Tate, the postmaster and jack-of-
all-trades who had written what, in essence, was the very
first Chamber of Commerce letter extolling the region's
attractions. Orville described Tate as "postmaster, farmer,
fisherman, and political boss of Kitty Hawk." He was a
typical Banker of his day—versatile, independent, self-
sufficient, rough-edged. He took Wilbur into his house as a
temporary guest, introduced him to the village, and, after
Orville's arrival on September 28, helped the two brothers
set up camp. Tate's wife loaned Wilbur the sewing machine
with which he trimmed down pieces of presewn fabric for
wing covers.

It must have been a curious sight: this well-dressed visitor
from Ohio assembling a glider with a seventeen-foot wing
span in the Tates' front yard, with an assist from Mrs. Tate's
sewing machine. Not a few Kitty Hawk residents must have
scoffed at the notion that any such contraption could actually fly. They must have also wondered aloud how long the
Wright brothers could survive autumn's steady diet of wind
and sand before taking flight back to Dayton.

The answer came soon enough. The brisk bite of Kitty Hawk's climate did not intimidate the Wright brothers. But their work was never easy. It was one thing to cite the area's average wind speed, which was conducive to the Wrights' glider experiments. But averages are made up of highs and lows. They also do not take into account that winds may shift directions without warning—from, say, soothing southwesterlies to the pugnacious winds of northeasters. In the first two weeks of October, the winds were either too light or too strong to attempt more than a single manned flight, which had to be aborted.

The brothers took to safer, unmanned flights to test their glider. They tethered the tips of the machine's wings with ropes and controlled it from the ground. As Wilbur later described the experiments to Chanute, "We spent quite a large portion of our time in testing the lift and drift of the machine in winds of different velocities, and with various loads."

But it was not simply the glider that was being tested. Accidents, delays, mishaps, and foul weather repeatedly conspired against the brothers. Over and over again, they persevered. On one afternoon they transported the glider to the hill just south of their camp. After flying the machine successfully, the brothers placed it on the ground to make adjustments. Suddenly the wind caught the glider's wing tip and flipped it twenty feet, causing severe damage. This setback marked one of several dark moments that would grip the brothers during their time on the Outer Banks. They seriously thought of giving up. "We dragged the pieces back to camp and began considering going home," Orville wrote his sister on October 14. "The next morning we had 'cheered up' some and began to think there was hope of repairing it."

By October 17, the much-improved glider was up and running again—but not before a forty-five-mile-an-hour northeaster had buried the machine in a mound of sand near their tent. It took the brothers half a morning to dig it out. The next day, they took the glider to some steeper dunes about a mile below their camp. But the wind died before they could fly with someone aboard.

The following day brought ideal flying conditions. For only the second time, Wilbur took the glider's controls. With two men running at the wing tips, the machine was repeatedly launched, and Wilbur brought it effortlessly back to earth. Some of the glides were as long as 300 or 400 feet.

In his impressive biography of the Wright brothers, Thomas Crouch summarized this phase of the adventure: "The exhilaration was incredible. Racing down the slope, holding his machine within five feet of the surface, Wilbur was traveling twice as fast at the end of a flight as at the beginning. He was flying, experiencing sensations known to only a handful of human beings." As Wilbur described it in a lecture a year later:

Although the hours and hours of practice we had hoped to obtain finally dwindled down to about two minutes, we were much pleased with the general results of the trip, for setting out as we did, with almost revolutionary theories on many points, and an entirely untried form of machine, we considered it quite a point to be able to return without having our pet theories completely knocked in the head by the hard logic of experience, and our own brains dashed out in the bargain.

October 19 was to be their last day of gliding that year; the winds were too light to continue. On October 23, the brothers' scheduled day of departure, they gave the glider one last fling into the sand flats, whence it came to rest. When the brothers returned the following year, they found the glider half-buried in sand. According to Crouch's account, the glider's last piece of wing disappeared in a ninety-three-mile-per-hour gale on July 25, 1901. Thus did the sands of the Outer Banks swallow another bit of history whole.

IF THE WRIGHT BROTHERS seemed well-suited by temperament to the Outer Banks' turbulent climate, it was not simply because their experiments depended crucially on the presence of a steady, strong breeze. They also appreciated the outdoors. After all, their chosen vocation entailed the use and repair of bicycles—machines which could only be used outside. They enjoyed hunting and fishing, and partook of both during slack times on the Outer Banks.

So it is not surprising to find in their letters and notes frequent references to their environment that transcend its relevance to their experiments. Particularly in Orville's letters to Katherine, his sister, there are descriptions of the bleak and beautiful scenery surrounding them and of some of the animals that populated it. As Orville wrote on October 14, 1900:

> This is a great country for fishing and hunting. The fish are so thick you see dozens of them whenever you look down into the water. The woods are filled with wild game. . . . At any time we look out the tent door we can see an eagle flapping its way over head, buzzards by the dozen . . . soaring over the hills and bay. . . . A mockingbird lives in a tree that overhangs our tent, and sings to us the whole day long. . . . I think he crows up especially early after every big storm to see whether we are still here; we often think of him in the night,

when the wind is shaking the top and sides of the tent till they sound like thunder, and wonder how he is faring and whether his nest can stand the storm. . . . The sunsets here are among the prettiest I've ever seen. The clouds light up in all colors in the background, with deep blue clouds of various shapes fringed with gold before. The moon rises in much the same style, and lights up this pile of sand almost like day. I read my watch at all hours of the night on moonless nights without the aid of any other light than that of the stars shining on the canvas of the tent.

In retrospect, it seems obvious: that two men who would test and discard prevailing theories of flight—who obeyed their own instincts and observations rather than the wisdom of the ages—would themselves be remarkably keen and sensitive observers of the world around them. Their eyes recorded not merely the geometry of the dunes and the velocity of the wind, but the aesthetics of both.

In the Wright brothers, as in so many other famous inventors and discoverers, aesthetic and scientific perspectives worked in tandem; mere sight was transformed into vision. They sought not to conquer the wind, but to harmonize with it. The secret, as they intuitively knew, was not to overpower the air, but to enlist it in their cause, to exercise *control*. The Wright brothers' understanding and appreciation of nature surely contributed to the complex set of factors that brought them breathtaking success where others had failed.

As one might imagine, the brothers' fascination with nature extended especially to birds. At Kitty Hawk, they were surrounded by birds. Indeed, the town's namesake, "kitty hawk," is a bird. The bird's name is not formally recognized by the American Ornithologists' Union (AOU), which dictates bird nomenclature—in North America, at least. One does not, however, have to go back far in the works of naturalists to find it.

The "kitty" part refers to the mewing noise the bird makes, and "hawk" to its swooping, predatory behavior. The beautiful little falcon, properly called a kestrel, regularly present on the Outer Banks, was often called a "kitty" or "killy" hawk. It does produce a high call, a bit too high-pitched and prolonged to be quite like a kitten's mew. On the Carolina coast, the name seems to have referred to gulls since Elizabethan times. As recently as twenty years ago, old-timers still called sea gulls "kitty hawks" at Hatteras and Ocracoke. Perhaps some still do.

A European counterpart of the laughing gull is called the mew, or sea mew, in onomatopoeic allusion to its call. The more pelagic gulls called "kittiwakes" are simply those who make the sound in a ship's wake out at sea. The predatory gulls called formally by the northern European names *skua* and *jaeger* were originally called "sea hawks" or "jiddy hawks" in English.

The enormous popularity of birdwatching has accelerated the rise of the AOU to its position of legal arbiter of official, standardized, "common" names for birds. While codifying the system is certainly not all bad, it has cost us some of the color of the country, some of the memorable and poetic names people used for the wild companions of their lives. So, today, when you ride a ferry or walk a pier, take a moment to watch the kitty hawk: rising—laughing gull indeed—then swooping, mewing like a kitten, to snatch a bit of bait. He is the Kitty Hawk of old, and if he did not provide Orville and Wilbur with technical specifications, he surely did give them inspiration.

That the flight of birds fascinated the Wright brothers is incontrovertible. Indeed, man's attempts to mimic the flight of birds dates back at least to Daedalus and the ancient Greeks. Otto Lilienthal, the German aeronaut whose writings and glidings helped kindle the Wrights' interest in flying, studied bird flight meticulously all his life and modeled his gliders' wings on bird and bat morphology. The title of one of his best-known books is *Bird Flight: the Basis of the Flying Art*. Like many would-be soarers before him, however, Lilienthal never unlocked the secret of flight, either by bird or man. He died in a gliding accident in 1896.

The Wrights' biographers note that the brothers frequently cited a book on ornithology whose reading in 1899 sharpened their interest in gliding. While the text was never identified, two books are possible candidates: James Bell Pettigrew's *Animal Locomotion, or Walking, Swimming and Flying, with a Dissertation on Aeronautics*, or translated portions of a book on bird flight by Frenchman Louis-Pierre Mouillard. Also highly influential in aeronautical circles at the time were the writings and experiments of Samuel

Langley, head of the Smithsonian Institution. Like Lilienthal, Langley had been fascinated by bird flight since his youth and attempted to apply his observations (unsuccessfully) to a variety of flying machines.

Author John Evangelist Walsh states that Wilbur Wright spent much of his free time in the two years after Lilienthal's death observing birds in flight. According to Walsh, Wilbur conducted much of his birdwatching in a location outside Dayton by lying on his back and observing birds through his field glasses.

But how much of this observation contributed directly to the Wrights' subsequent breakthroughs? What did Wilbur and Orville see in the flight of birds that Lilienthal, Langley, Chanute, and a bevy of other scientists and engineers did not see? "We could not understand that there was anything about a bird that could not be built on a larger scale and used by man," Orville once stated. "If the bird's wings would sustain it in the air without the use of any muscular effort, we did not see why man could not be sustained by the same means."

Wilbur reported that his study of the flight of pigeons led him to the concept of lateral control in his kites. Lateral control describes the way in which a bird lifts or lowers the tips of its wings to effect a turn or maintain control. Similarly, Wilbur figured, the wings of a glider could be twisted in opposite directions to make a turn—wingwarping, as the technique was dubbed. It was a pivotal discovery in the Wrights' early thinking about how gliders could use the air—rather than some human factor, such as shifting weight—to maintain control.

With their large assortment of soaring and gliding birds, from eagle to hawk to buzzard, the Outer Banks became a birdwatchers' paradise for Wilbur and Orville. "Kitty Hawk

is a splendid place to observe soaring flight," Wilbur wrote
Octave Chanute in November 1900. "I think at least a
hundred buzzards, eagles, ospreys, and hawks made their
home within a half mile of our camp. We were enabled to
make a number of observations and settle conclusively to
our minds some points which have been much disputed
among writers on the soaring problem."

Detailed references to birds are sprinkled throughout the
Wright brothers' diary entries. "The bird certainly twists its
wing tips so that the wind strikes one wing on top and the
other on its lower side," Wilbur wrote of the pigeon, "thus
by force changing the bird's lateral position." And again: "A
bird sailing quartering to the wind seems to always present
its wings at a positive angle, although propulsion in such
position seems unaccountable."

The observations cover a variety of species in a variety of
situations. "At 8:30 saw buzzards soaring over sand hills,"
Orville wrote on September 16, 1902. "Conditions seemed
to be such that they were not able to soar over plains but
took to hills where they had considerable trouble in gaining
altitudes of more than 50 to 75 feet above top of large hill.
We watched them with field glasses at a distance of 1,200
ft."

The importance of these observations, some of them quite
technical and detailed, was confirmed in Wilbur's lengthy
address to the Western Society of Engineers in June 1903.
Wilbur devoted the final section of his speech to birds and
flight, and concluded with this comment, worthy of Daedalus himself: "There is no question in my mind that men can
build wings having as little or less relative resistance than that
of the best soaring birds. The bird's wings are undoubtedly
very well designed indeed, but it is not any extraordinary
efficiency that strikes with astonishment but rather the

marvelous skill with which they are used. . . . The soaring problem is apparently not so much one of better wings as of better operators."

But just when it seemed safe to conclude that the Wrights' fascination with birds did contribute to their technical understanding of flight, Orville dismissed the suggestion years later. He argued that their observations of birds in flight tended to confirm ideas they already had, or decisions they had already made. "Learning the secret of flight from a bird was a good deal like learning the secret of magic from a magician," he wrote. "After you know the trick and what to look for, you can see things you didn't notice when you did not know exactly what to look for."

Orville's explanation seems plausible but also somewhat pat, since he was speaking from the perspective of hindsight. While it would be simplistic to suggest that birds taught the Wright brothers how to fly, our feathered friends clearly warrant more than a footnote in the annals of aviation.

IF THE ECOLOGY of the Outer Banks ideally suited the Wright brothers' purposes, it also periodically conspired against them. For example, while the winds of Kill Devil Hills often exceeded the brothers' expectations, those stiff breezes could also be accompanied by atrocious weather in which neither man nor beast would attempt to fly.

The number of days on which the Wrights could actually test their machines was therefore smaller than they expected. As Wilbur Wright explained in September 1901, only gusts of thirty miles per hour gave their 1900 flying machine the angle of incidence it needed to glide. "As winds of 30 miles per hour are not plentiful on clear days," he said, "it was at once evident that our plan of practicing by the hour, day after day, would have to be postponed."

During their first autumn on the Outer Banks, foul weather plagued them from the moment Wilbur began bailing out Israel Perry's leaky vessel. As Orville wrote Katherine on October 18, 1900:

> Our nights in Kitty Hawk are interesting and, were there not so many of them, not unpleasant. . . . About two or three nights a week we have to crawl up at ten or eleven o'clock to hold the tent down. When one of these 45-mile nor'easters strikes us, . . . there is little sleep in our camp. . . . The wind shaking the roof and sides of the tent sounds exactly like thunder. When we crawl out of the tent to fix things outside the sand fairly blinds us. It blows across the ground in clouds. We certainly can't complain of the place. We came down here for wind and sand, and we have got them.

These blustery nights were cold, too. "We each of us have two blankets, but almost freeze every night," Orville wrote in the same letter. "The wind blows in on my head, and I pull the blankets up over my head, when [sic] my feet freeze, and I reverse the process."

But, having snared the brothers with their beauty and their breezes, the Outer Banks reserved their harshest blow for the second time around. The Wrights returned to Kitty Hawk earlier in 1901, in July, only to find that what islanders called "the greatest storm in the history of the place" had just passed through. Anemometers had stopped working at ninety-three miles an hour, the weather station reported. The storm broke a seven-week drought and triggered rains for a week. "This has delayed us beyond expectation . . . ," Wilbur wrote Katherine on July 26.

A week of rain on the Outer Banks also guarantees something else: a plague of mosquitos rivaling the Old Testament locusts. It is ironic that one of nature's smallest airborne creatures almost succeeded in driving off the inventors of man's first flying machine. As Wilbur wrote Katherine, "...it seems nature has been in a conspiracy with our enemy, the mosquito."

While the brothers had been besieged with northeasters during the previous year, they longed for one now, to blow away the clouds of mosquitos feasting on them. "The agonies of typhoid fever with its attending starvation are as nothing in comparison," Orville wrote Katherine. "But there was no escape. The sand and grass and trees and hills and everything was fairly covered with them. They chewed us clear through our underwear and socks. Lumps began swelling up all over my body like hen's eggs."

This misery carried through the next day and into the night. It prevented any work on the new glider the brothers were building. That night the men attempted to elude their adversaries by building mosquito netting and putting their cots outside on the sand. But, as Orville reported, they were quickly routed. "The tops of the canopies were covered with mosquitos till there was hardly standing room for another one; the buzzing was like the buzzing of a mighty buzz saw," he wrote. "...Affairs had now become so desperate that it began to look as if camp would have to be abandoned or we perish in the attempt to maintain it."

The next morning, however, the brothers and their visitors, including their friend George Spratt, dragged in old tree stumps and lit fires around the camp. The mosquitos gradually began to disappear, and the wind to rise. There would be more gliding than scratching, at last.

Even so, the 1901 experiments on the new machine proved both dangerous and disappointing. By the time the brothers departed Kitty Hawk on August 22, they were thoroughly discouraged. "When we left Kitty Hawk at the end of 1901," Wilbur remembered much later, " . . . we doubted that we would ever resume our experiments. . . . At this time I made the prediction that men would sometime fly, but that it would not be within our lifetime."

That was one of the Wrights' few inaccurate predictions.

Over the winter of 1901–02, they experimented with wing designs and a wind tunnel. They returned to Kitty Hawk in September 1902, with a renewed sense of purpose and vigor. Their winter work paid off: they were quickly aloft in their new, improved glider, which occasionally soared more than 500 feet away.

It was during this period that the Wrights sealed their friendship with George Spratt, who had joined them briefly the previous year. Spratt, a flying enthusiast and physician from Philadelphia, provided good company. Orville and Wilbur especially appreciated Spratt's detailed knowledge of the flora and fauna of the area. "Spratt is a fine fellow to be with in the woods," Orville wrote Katherine, "because he knows every bird, or bug, or plant you are likely to run across." Spratt's love of the outdoors appeared to be one of the main reasons the Wrights took a liking to him: a privilege granted to very few people in this intense, trying period of their lives.

The weather, as always, continued to work both for and against the brothers in no particular pattern. "Many days were lost on account of rain," Wilbur later told an engineers' society. "Still more were lost on account of light winds." He explained that only when the wind reached twenty miles an hour did gliding become "real sport," because the wind helped them carry the machine back uphill. Toward the end of this five-week sojourn at Kitty Hawk, the brothers recorded more than 250 glides in two days. One glide succeeded in a thirty-mile-per-hour wind, a record. A record distance of 622.5 feet was achieved in another, along with a record time aloft of twenty-six seconds. The brothers had solved most of the design problems that had plagued them. All they needed now was an engine.

By 1903, the Wright brothers' return to their old haunts in Kitty Hawk had assumed a certain ritual. The first task, of course, was to check the previous year's building for damage and repair it. In 1902, for example, wind had so eroded their building that it had sunk two feet into the sand and now filled with twenty inches of water whenever it rained. The following year, they found the building "several feet nearer the ocean" than the previous year, and about a foot lower in several places. No matter: they erected another building next to it, in which to assemble and house their new, powered machine.

Another part of Wrights' ritual of return was to hear from the Bankers just how foul the weather had been during the intervening months. The year 1903 was no exception: Kitty Hawk residents reported that the storms had been of "unprecedented severity," according to Orville, and that "the mosquitoes were so thick that they turned day into night, and the lightning so terrible that it turned night into day." Still, he concluded, "these sturdy Kitty Hawkers have survived it all, and are still here to welcome us among them."

For obvious reasons, 1903 is remembered as the Wright brothers' most celebrated time on the Outer Banks. It is appropriately memorialized in every child's history book. What is not often recorded, other than the years of work and mishaps that had preceded it, is that the brothers had to live through yet another terrifying storm before they won their wonderful, Daedalean moment in the sun. Indeed, their dreams of flight were almost drowned in sheets of rain.

In his letter to Katherine on October 18, 1903, Wilbur called this storm a "cyclone," but in today's parlance it was a hurricane—with winds circulating in a counterclockwise direction. Comparing it to the deluge of biblical times, Wilbur reported that the storm stalled off the coast and "backed" into them seven times over a four-day period. On the storm's first night, with fifty-mile-an-hour winds, the brothers fully expected their new building to collapse, with them in it. The old building flooded.

By four o'clock the next afternoon, after the brothers had furiously braced the new building with hammers and nails, the wind velocity hit seventy-five miles an hour—hurricane force. When the building's tar-paper roof began to give way, Orville headed out into the teeth of the storm with a ladder. As a bemused Wilbur later described his brother's progress, "he began by walking backwards about 50 feet." When Orville mounted the ladder, now with Wilbur's assistance, the wind blew so hard that it caught Orville's coat and "folded it back over his head." Since his hammer and nails were in his coat pockets, he couldn't reach them and had to descend the ladder. Finally, the roof was repaired, but not before the brothers were soaked to the bone and Orville had thoroughly hammered his thumb.

"The wind and rain continued through the night," Wilbur ruefully recorded, "but we took the advice of the Oberlin coach, 'Cheer up, boys, there is no hope.'" By morning the floor of their building was largely underwater, though the kitchen and dining room were spared. The storm continued for two more days, Wilbur continued, "but by Monday it had reared up so much that it finally fell over on its back and lay quiet."

And what did the Bankers say? That this blow was the most persistent they had seen and one of the strongest. Five ships were reportedly washed ashore between Kill Devil Hills and Cape Henry. In truth, considering the Banks' long history of violent storms, the Wrights were fortunate that they did not become the Lost Aviators of their day.

Still, the brothers continued their work. On clear, windy days they flew an unmanned glider. When the weather forced them indoors, which was often, they labored over their powered machine. The bad weather mirrored seemingly insurmountable mechanical difficulties within. When Orville wrote in his diary on November 20, "Day closes in deep gloom," he wasn't simply referring to the weather.

Perhaps the irrepressible Chanute was right: "He seems to think we are pursued by a blind fate from which we are unable to escape," Orville wrote home. Through November's bitter cold, they hung on. An uncooperative engine was tested and repaired. Finally, on December 14, all was ready. With Wilbur at the controls, the new machine took off all too abruptly: it rose fifteen feet in the air and sailed sixty feet from the end of the launching track before landing and spinning around in the sand. Though the machine was damaged, the brothers knew that, except for Wilbur's unsteadiness at the controls, they had won. "There is now no question of final success," Wilbur exulted later that day.

As usual, Wilbur was right. After making repairs and waiting for a stiff north wind of between twenty and twenty-five miles per hour, the brothers pushed their machine out of the hangar on Thursday, December 17. They readied it on the launch rail they had laid in the sand. Orville and Wilbur shook hands. At 10:35 A.M., with Orville at the controls (it was his turn), and in Orville's matter-of-fact words, "The machine lifted from the truck just as it was entering on the fourth rail." Twelve jerky seconds and 120 feet later, the machine skidded to a halt. Triumph for man in wind and sand.

After a few repairs, Wilbur made a similar flight, followed by Orville and then Wilbur again. On this fourth and last flight, unequivocal release from earth's bounds was truly achieved: Wilbur flew 852 feet in fifty-nine seconds. Nearly as remarkable was that Orville's first flight was duly captured on film, thanks to John Daniels of the Kill Devil Life Saving Station, who snapped the picture just as Orville lifted off the track. Biographer Tom Crouch describes the shot, unarguably, as "one of the most famous photographs in history."

As a symbol of the Wrights' splendid accomplishment, the photo radiates meaning. Appropriately, the two brothers are alone with their machine, dwarfed by it, with no assistance other than that provided by the wind and sand and their own ingenuity. If humans and their environment ever coexisted in near-perfect harmony, this photograph captures that ideal state. Unlike the noisy labors that preceded it, the birth of flight appears serene and simple.

But nature was not done with the Wright brothers yet. It was ready to deliver one more capricious blow to them. After the fourth flight that day, Orville and Wilbur contemplated a much longer excursion by air—perhaps up to the Kitty Hawk weather station. But as they discussed this trip a sudden gust of wind flipped up one wing tip, causing the machine to roll over backward, with John Daniels clinging desperately to it. Daniels was not injured, but the machine was destroyed. The same breeze that had liberated the brothers now cavalierly swatted their work away, like flotsam on the beach. There would be no more flying this year—courtesy, as always, of the wind.

THE WRIGHT BROTHERS did not return to Kitty Hawk for another five years. Their next series of flights took place at Huffman Prairie, a large cow pasture situated only

On December 17, 1903, at Kitty Hawk, Orville Wright became the first human successfully to sustain flight in a powered, heavier-than-air machine. Biographer Tom Crouch calls this "one of the most famous photographs in history."
© Smithsonian Institution, photo #A-26767-B-2

eight miles east of Dayton. They opted against more long, expensive trips to the Outer Banks, and their machine no longer needed an assist from the Banks' dunes and wind.

By 1908, however, the brothers were ready to test a new, more powerful, two-seater machine. They chose Kitty Hawk again, partly to elude the inquiring eyes of the press. As Tom Crouch describes the bittersweet scene the brothers encountered on their arrival in April of that year: "There was not much left. The side walls and south end of the original hangar were still standing, but the roof and north side had collapsed. The floor lay beneath a foot of sand and debris. The 'new' building had vanished completely, victim of a recent storm.... The skeleton of the 1902 wing protruded from a small dune just east of the original hangar.... An odd assortment of bits and pieces—ribs, spar sections, the cradle of the 1903 machine—littered the surface."

Within a few weeks, however, the camp was up and running and the brothers were ready to fly. Despite their best efforts to maintain privacy, members of the press swarmed on the Banks. Time after time, the brothers took their machine aloft, for flights exceeding 2,200 feet. Fearful that the Wrights would suspend their flying if they detected their presence, a group of reporters trekked to a hiding place in nearby woods—where, in the words of one reporter, they were "devoured by ticks and mosquitoes, startled occasionally by the beady eyes of a snake and at times drenched by a heavy rain." Orville and Wilbur would have enjoyed this moment, had they known about it. The reporters were rewarded for their discomfort. The airplane soon zoomed over them at an estimated forty miles per hour, turning gracefully in the air.

But not everything proceeded according to script. On May 14 Wilbur commenced a long flight, the third of the day. The machine disappeared over the dune, whereupon Wilbur lost control and crashed it, at forty-one miles per hour, into the sand. Although Wilbur suffered only minor injuries, the machine was once again destroyed. The crash made national headlines, and there would be no more flights at Kitty Hawk that year.

Wilbur, who died in 1912, would never again return to Kitty Hawk. But Orville did make two repeat engagements. The first was in October and November 1911, when he came to test a new glider. Orville made over ninety glides during this period, one of which set a world record for soaring flight—nine minutes, forty-five seconds. The record stood for a decade.

In November 1932, Orville returned to Kitty Hawk for a different purpose—to attend the dedication of a national monument to the brothers' historic achievements. The well-known monument, constructed of Mount Airy granite, is a sixty-foot-high pylon bearing chiseled wings on its sides. But like every other human activity on the Outer Banks, building the structure was not easy. First Kill Devil Hill itself had to be stabilized. Otherwise, the winds would continue to push the hill toward the sound at the rate of twenty feet a year.

The task of stabilizing the hill fell to Captain John A. Gilman of the Army Quartermaster Corps. First Gilman built a fence around the hill to keep grazing animals out. Then he covered the lower northeast slope of the hill with a two-inch layer of mulch and planted hearty grasses in it. As the grasses rooted, he continued the plantings to the dune's peak and down the other side. The project worked. The hill's march to the sound was halted, and the monument was safely built. Gilman's was one of the most successful stabilization endeavors anywhere on the Banks. Today the hill is covered with vegetation.

The monument's presence soon opened Kitty Hawk, Nags Head, and the surrounding area to some of the most rapid commercial and residential growth anywhere in North Carolina. Tourists from around the world flock to see the monument and accompanying National Park Service museum. The surrounding rows of houses, motels, and restaurants obscure the Wright brothers' real achievements, which are forever fused with the sand and wind and sky they loved.

IN DECEMBER 1903, Orville Wright was thirty-two, Wilbur thirty-six. On the twenty-ninth of that month, a young American born at Nice, France, celebrated his twenty-fourth birthday. His career was already well begun. He had galloped with Teddy Roosevelt's Rough Riders in the liberation of Cuba. He had indeed carried a big stick, but he would never walk softly. His imprint would become indelible. It will remain so as long as humans attempt travel in the sky.

Billy Mitchell was to become an undisputed hero in the First World War, a beloved leader of men in action, thought, and word. He was to be the youngest member of the U.S. Army general staff, a brigadier general, and the father of the Air Force. In the words of historian Roger Burlingame: "He told us that there had been a revolution in the world, a revolution born in America on the sands of Kitty Hawk."

Like Wilbur and Orville, General Mitchell was an avid hunter and naturalist. He described spiders using silk for "ballooning" in the Great Dismal Swamp, hunted tigers in India, and chronicled the fauna of the remote Typhoon Islands in the Philippines. Beginning about 1921, he frequented the Outer Banks, landing on the beach in his De Havilland. He hunted and fished, told war stories, and made friends.

In his running battle with the U.S. Navy (and ultimately the army, too), Billy Mitchell drew the world's attention to the Outer Banks in yet another context. To demonstrate his point that warships were indeed vulnerable to skilled aerial attack, Mitchell chose a site just off Cape Point for an exercise in ship-sinking. His choice was dictated by the need to find a water depth of at least fifty fathoms, so that the sunken ships would not pose navigational hazards.

In his famous and successful trials of 1921, Mitchell sunk the German "war prizes" *Frankfurt* and *Ostfriesland*. But in order to accomplish the ships' demise, Mitchell was forced to fly more than a hundred miles, from Langley, Virginia.

Had he been able to fly from Hatteras, the distance would have been cut to thirty miles.

In the summer of 1923 Mitchell built an airstrip at Hatteras preparatory to sinking the decommissioned battleships *Virginia* and *New Jersey*. The War Department changed the prerequisites for the tests at the last moment, requiring a bombing run from 10,000 feet. Mitchell's crews worked straight through the Labor Day weekend to prepare six planes to climb to this height with full bomb racks. Five of these completed the run on September 5, flying from Langley under the command of Lieutenant Charles Austin. They each carried four 600-pounders.

They departed Langley at six in the morning and landed, after their attack on the *New Jersey*, at Hatteras at 8:40 A.M. It took them most of their trip south to gain the required elevation, and they could not carry sufficient fuel to make the return trip. It is not specifically noted in our sources, but apparently this run was less than a glowing success.

With Mitchell now at Hatteras, the second phase began. This involved seven planes commanded by Captain Lloyd Harvey, each carrying one 2,000-pounder. Captain Harvey had moved his flight down to Hatteras on Tuesday, September 4. Murphy's Law prevailed, however, and most of these big bombs failed to inflict maximum damage because they stuck in their racks or the fuses either failed entirely or were improperly timed, detonating the bombs at the wrong depths.

Mitchell next sent up Lieutenant Harrison Crocker with seven planes fitted with 1,100-pounders, and told them to drop from 6,000 feet on the *Virginia*. The ship sank in thirty minutes. Heartened but running out of fuel, Mitchell finally drained all the tanks available to send two planes out again with two 1,100-pounders each to finish the *New Jersey*. Of these four bombs, the first was a clean miss and the second a dud. The third, however, was a solid hit and the *New Jersey* immediately turned over.

Thus the successful trials of September 1923 must be credited to the men flying from the little packed-sand strip at Hatteras. And the Outer Banks were critical in the career of the man who has proven to be the most powerful aeronautical prophet of the twentieth century. Surely no one person gambled so much, lost so much, won so much, saw so clairvoyantly the future of flight, and spanned this century's developments so effectively as Billy Mitchell. Airmail, passenger airlines, and jet fighters were all predicted by Mitchell, and all proclaimed preposterous by his noisy detractors, who seemed to prevail at the time.

Mitchell's most celebrated prophecy was apparently first articulated in 1923. He said the entire U.S. Pacific Fleet was potentially helpless, and that planes could fly from flat-topped ocean vessels at sea. He said those planes could fly from east of their targets—coming west out of a blinding, rising sun—and could drop not just bombs, but torpedos, on our ships at anchor in Pearl Harbor, Hawaii. He said they could easily destroy our meager air defenses on their runways.

Admiral William A. Moffett gave the official pronouncement on those maverick views, saying, "their author is of unsound mind and is suffering from delusions." Eighteen years later, on December 7, 1941, that attack came and America entered World War II. Dying in 1936, Billy Mitchell said, "everything I have said about airplanes and battleships will be verified in the war. . . which will be upon us in less than five years." He was a man who died with total faith in his ideas, many of which were tested at the Outer Banks.

≈

POETS AND PHILOSOPHERS have long been fascinated with the idea of flight. One poet who made his way briefly to the Outer Banks, and who regarded that experience as a turning point in his life, was Robert Frost. Frost later wrote a poem about it: "Kitty Hawk" mixes together personal experience, observations on the Wright brothers, and a paean to Western philosophy and civilization.

Frost literally stumbled on the Outer Banks by accident. In November 1894, spurned by his fiancée, Elinor White (who would eventually become his wife), Frost abruptly left his Massachusetts home for the Dismal Swamp along the Virginia–North Carolina border. Frost's biographer, Lawrance Thompson, wrote that Frost, only twenty at the time, seriously contemplated suicide.

Once in the Dismal Swamp, Frost wandered until he came upon an old boat plying the Dismal Swamp Canal to Elizabeth City. The boaters intended to take a party of hunters across Albemarle Sound to a place Frost had never heard of—Nags Head, on the Outer Banks. Frost boarded the boat and slept much of the way. Befriended by the convivial hunters, young Rob joined them for drink and food at a small Nags Head hotel.

Later in the evening Frost slipped away, walking past Jockey's Ridge, over the dunes, and down to the ocean. There he encountered a member of the local lifesaving crew, with whom he walked and exchanged conversation, some of it about the area's stories and legends. Frost apparently spent the next day or two walking the area, ranging as far south as Oregon Inlet. He then retraced his steps to Elizabeth City and rode the railroad to Baltimore, sleeping in boxcars and living with hobos. He returned to Massachusetts, somewhat chastened but poetically inspired. On this impressionable, romantic New England writer, the trip had left an indelible mark.

In later life Frost often talked of the experience, though details sometimes varied in the telling. What he did not know at the time is that two of his subsequent heroes, Wilbur and Orville Wright, would soon retrace his steps to Elizabeth City and the Outer Banks. Frost once compared the Wrights to Christopher Columbus, an explorer who had "the faith of an idea." Except for Wilbur and Orville, he said, no other pioneers could match Columbus's place in history. Frost also said, in conversation with a friend, that "Anybody who knows even a kindergarten course in my poetry knows that I've been interested in flying ever since Kitty Hawk gave us success under the Wright brothers."

"Kitty Hawk" was first published in 1957 and dedicated to Huntington Cairns, a lawyer and writer who owned a second home near Kitty Hawk, and who persuaded Frost to revisit his youthful haunts there in the summer of 1953. Among other themes, the poem expresses Frost's admiration for the Wrights' achievement. The poem suggests that Frost met Orville—a meeting which in fact did take place, in Washington during the 1930s:

> Once I told the Master,
> Later when we met,
> I'd been here one night
> As a young Alastor
> When the scene was set
> For some kind of flight
> Long before he flew it.

Using the sands of Kitty Hawk as his metaphorical launching pad, and the Wright brothers as his heroes, Frost makes his own flight of the imagination. It is a lofty, somewhat arrogant excursion through Western philosophy and religion—a defense of the West's grounding in material, as well as spiritual, values. From such material pursuits, Frost argues, Westerners have pushed the limits of knowledge, and have turned that knowledge into action. Man is the only thinking being in our solar system, he reminds: "Let's keep starring man / In the royal role."

It is the role of man also to keep stirring the pot of action, and to impose his own theories and meaning on the "vague design" of nature. Whether the Wright brothers would have fully subscribed to Frost's expansive view of man's glorious future on earth is doubtful. And whatever they would have thought in private about such matters, they were not likely to proclaim their views publicly from the rooftops. No matter. With a deceptively deferential doff of their hats, Orville and Wilbur no doubt would have allowed Frost the final say:

> God of the machine,
> Peregrine machine,
> Some of us think is Satan,
> Unto you the thanks
> For this token flight,
> Thanks to you and thanks
> To the brothers Wright
> Once considered cranks
> Like Darius Green
> In their home town, Dayton.

Chapter Eight

Convergences

It shouldn't take him by surprise, he'd always known
both sides: that the salt marsh is the salt marsh, the
sea is the sea, the sky is the sky . . . and that the
land washes into the salt creek, the salt creek
into the sea, the sea into every sea, and
everything in the sea dissolves.
John Casey, 1989

The U.S. outer continental shelf is one of the areas
most likely to contain undiscovered oil and gas in
large quantities—perhaps as much as one-half
of this nation's future petroleum.
American Petroleum Institute, 1986

T WO HUNDRED MILLION years ago, in a dim and quiet time of shallow seas and falling sea level—a time quite unlike our own—in the most immediate, pragmatic, polemical, legal, and financial way, our future began. That was about the time the first toad croaked in a remote freshwater marsh. It was some 140 million years before the first songbird warbled. And it was at least 199.8 million years before the first hirsute human ever stood before the motley assemblage of his tribe and promised to construct some wall of rocks at someone else's expense, for some pronounced benefit to his listeners (and clearly perceived, but unstated, benefit to himself).

Our future—in the closing decade of this century and for decades foreseeable into the next—is in largest measure tied to the demography of organisms that lived and died in that silent setting all those millions of years ago. From their teeming numbers in the past comes "rock oil," rendered in scientific Latinese as *petroleum*. Generally speaking, marine invertebrate animals cooked down—quite literally, with the heat and pressure of their fossil interment—into oils and tars, while plants have been rendered as natural gas. Yale geologists Richard Foster Flint and Brian Skinner say of petroleum's origin: "The raw material consists mainly of microscopic marine organisms, mostly plants, living in multitudes at and near the sea surface. Measurements show that the sea grows at least 31,500 kg of organic matter per square kilometer per year, and the most productive inshore waters grow as much as six times this amount."

There are about 35 tons in 31,500 kg, so six times that— 210 tons—can be produced per year on less than two-fifths of a square mile (about a square kilometer) of sounds, embayments, and estuaries. That number dwarfs agricultural production on the finest of earth's croplands. What sea water did in the past it does today, and can continue to do in the future, if we become much more zealous than we have been in preventing the sea's degradation.

It is instantaneously apparent that in coastal Carolina the sea, the sounds, the embayments, and the estuaries mean everything for the future: economically, biologically, geologically. That the sea of 200 million years ago is every bit as important to the people of Akron, Ohio; Missoula, Montana; and Chengdu, China, is somehow less apparent, but no less true. Oil is not only present every time one starts a car, in every ounce of fuel, in every dollop of lubricant, and every drop of paint. It is in every synthetic chemical, from nylon to the ink in a pen—and the plastic shaft of the pen as well: petroleum is Everyman's instant proof that everything is connected to everything else.

If petroleum, in all its solid, liquid, and gaseous states, with all its various products and byproducts, is considered one thing, no one thing is arguably more important to maintaining our modern life; its various forms have become as fundamental to our daily living as oxygen and water are. And it happens to be right there: right off the Outer Banks. There for the tapping. There for the risks. The only questions are: Precisely where is it? In what form? How much? Is it worth drilling for?

". . . The unfortunate reality is that, short of eliminating oil transportation at sea entirely, there is no perfect solution to offshore oil spills. *It is certain that oil spills will occur again.*" So stated the U.S. Congress's Office of Technology Assessment (the emphasis is theirs) following the Exxon *Valdez* disaster at Prince William Sound, far away in Alaska. As far away as your television set, your gas pump, your furnace, and every breath of air you have inhaled since the very first spill.

In September 1988, Mobil Oil Company announced its intention to submit a plan to explore for natural gas reserves in the waters off Cape Hatteras. If Mobil has its way, a 534-foot drillship will set up shop over Manteo Block 467. Block 467 is a patch of ocean floor that lies thirty-eight miles due east of the Hatteras village of Salvo, or forty-five miles east and north of Cape Point. Manteo Block 467 has become one of the hottest and most disputed pieces of real estate in North America.

Mobil has given its proposed drilling site the rather alluring name of "Manteo Prospect." The Manteo Prospect is considered one of the most likely spots anywhere along the East Coast for a major gas field. The site is located on an ancient, buried reef that existed along the margin of the Atlantic's western continental shelf more than 140 million years ago. The porous limestones that form the reef provide an ideal reservoir for natural gas percolating up from the deeper Jurassic beds below. In addition, the Manteo Prospect is located on a large, structural ridge called the Norfolk Arch. Since the arch is higher than the surrounding terrain, it is a likely place for hydrocarbons to collect. Oil company geologists say the reef could yield 5 trillion cubic feet of natural gas. That would make it one of the largest known gas fields in the world and the largest discovery of natural gas in the United States since Prudhoe Bay, Alaska, in 1968.

The target area is approximately thirty miles long and six miles wide. Water depths in the region range from 600 feet to more than 6,000 feet. The oil companies have already paid the federal government nearly $300 million for leasing rights, but the oil slump of the early 1980s prevented them from testing their hunch until the end of the decade.

The Manteo Prospect is in 2,690 feet of water—well within the record water depth for offshore wells of 7,638 feet. Recent industry advances have made offshore drilling much more routine in deep and turbulent waters; the North Sea project testifies to that. The industry maintains that 107 exploration wells have been drilled in waters deeper than the proposed Mobil well. The well itself is expected to go some 14,000 feet below the ocean floor and would be thirty inches in diameter at its widest point. The proposed drilling would occur over a four-month period and would cost about $25 million. All told, Mobil and its partners can expect to spend up to $1 billion on their high-risk investment.

But just how high-risk? Mobil estimates that the chances of finding natural gas are only one in ten. Some geologists believe the chances are actually more remote than that. The drilling record in the offshore Atlantic is not very encouraging. Of the fifty-one wells drilled in the North, Middle, and South Atlantic waters, only five have produced measurable quantities of hydrocarbons—and none in economically viable amounts. Success has been much greater off the coast of Canada, where Mobil twice discovered hydrocarbons in 1979. In September 1990, Mobil Oil Canada signed a $4.5 billion agreement to begin drilling for offshore oil in the huge Hibernia oil field about 200 miles southeast of St. John's, Newfoundland.

The chances of discovering oil in the proposed Manteo Prospect are deemed much worse—only one in a hundred. But such long odds have not discouraged oil companies from exploring North Carolina waters. In 1950, Exxon drilled an exploratory well off Cape Hatteras to a depth of 10,000 feet, and found nothing. More than fifty wells have been drilled both onshore and in coastal waters controlled by the state. Mobil drilled three wells in Dare and Hyde counties in 1965. Neither oil nor gas was found in any of these operations. Still, modern oil explorers persist. Earlier drillings seem only to have strengthened their resolve to find gas off the North Carolina coast.

Most residents of the Banks do not appreciate the thought of drilling off their precious and largely pristine coast. They see very few advantages, most of them ambiguous at best, and quite a few risks. The economic argument does not move them much. Public services such as roads, schools, and water supplies are already sorely taxed. So the additional jobs and capital spending that an offshore rig would bring do not excite the local Chamber of Commerce; for them, any benefits would likely be canceled out by having to provide more services.

To the extent that the rig might mean oil spills or tanker accidents, local officials think of the negative impact on the region's main economic staple: tourism. Then there is the aesthetic argument. North Carolina's coast has so far avoided many of the trappings of progress that have besmirched coastal areas to the north and south. Why invite any activity that could threaten the very remoteness and beauty that make the Outer Banks a national treasure? Why even entertain visions of an oil or gas rig on the Cape Hatteras horizon?

Mobil counters that it expects to find only gas, not oil. If that is true, then the environmental impact would be relatively small. Natural gas is considered the prime fuel of the twenty-first century. The United States acquires 25 percent of its natural gas from offshore wells, and a White House energy study predicted that at least 60 percent of the fossil fuels recovered offshore in the next century will be gas.

Natural gas is a clean fuel that can be piped to shore. Gas pipelines from Block 467, however, would most probably have to be routed well south of the Banks since Dare County, which includes Hatteras Island, has banned pipelines, refineries, and other accoutrements of the petroleum industry. The gas would be processed in a small onshore facility. Gas does not require tankers for transport and, when spilled, does not blanket the ocean. A single gas platform would not be visible from the Cape Hatteras Lighthouse and would represent a minor intrusion on a vast seascape.

Comparison may be made to Alabama, which opened its first offshore rig in 1988—a natural gas well drilled by Mobil. Alabama officials say that while the state has profited handsomely from the taxes and royalties involved, the area has not experienced dramatic new growth. Ancillary industries that have sprung up because of the rig had to be lured there by the state. The rig itself employs only about 30 to 40 people, and another 100 work at a small, onshore gas refinery. Alabama officials have reported no quantified environmental damage during the years that Mobil has explored and worked there. But there have been problems: ten years ago, after being fined $2 million for pouring untreated sludge and toxic drilling "mud" into Mobile Bay, the company agreed to comply with state rules and take all discharges to land to be treated and disposed of. That only begs the perennial question: how does one safely dispose of toxic wastes?

But Bankers are concerned that whatever boomlet a single gas discovery might elicit will not end there. First, no guarantee exists that oil will not be discovered instead of, or along with, gas. Mobil acknowledges that oil could be found. Before it is allowed to explore, the company will be required to develop an oil spill contingency plan just in case. And even if only gas is discovered, surely Mobil or its partners will drill for more gas and oil along the same reef. It is not beyond the realm of possibility to imagine rigs sprouting up all along the coast.

Geologists like Stan Riggs caution against getting too carried away with immediate visions of platforms and pipes. The odds against a discovery are good. It can be argued that it would be better to allow the drilling on the theory that if gas is not discovered, the oil companies will fold up their rigs and try elsewhere. The more the ante is upped, and the more the companies are required to spend fighting off opposition, the more determined they will become to justify a return on their mounting investment. Some say it is better to let them drill quietly, and fail, than to oppose them and cause them to redouble their efforts—and win. But what if they drill quietly now, and win?

There is a larger picture. It comes into sharpest focus in the waters today overlying that ancient reef which Mobil hopes has captured the hydrocarbon remnants of all those uncounted Jurassic marine organisms. You will recall that the Gulf Stream, that hose-pipe of warm water, rushes up the outer continental shelf and turns sharply east off Cape Hatteras. The Stream creates little gyres of surface water that curl away regularly along it, into the bights between the elbow capes of North America. A big gyre curls back across the North Atlantic to contribute its water, now much cooled by drifting icebergs, to the Labrador Current sweeping south along the northwestern edge of the Atlantic, the northeastern rim of the continent. The grandest gyre of all proceeds on around the North Atlantic, a vast clockwise circulation that eventually brings the water back to its source off equatorial Africa, where the trade winds leap forward to push it westward again.

Somewhere off Hatteras (the exact position is nebulous and changeable), the descendants of the Labrador Current, coming south, meet the Stream, going northeast, carrying monstrous mats of *Sargassum*—the seaweed that floats eternally in the calm center of the grand gyre of the North Atlantic—and all that blends with the nutrient-rich outwash of the continent, funneling through the inlets of the Outer Banks. This zone of confluences, roughly overlying the Manteo Prospect, is called (incongruously enough) "the Point."

Quite apart from the threat of an oil spill, opponents of drilling have focused their attention on the potential impact petroleum could have on the rich fishing beds of the Point, an area considered one of the best on the East Coast for commercial and recreational fishing. Just as Buxton Woods serves as a north-south meeting ground of species of plants and animals on shore, so the Point provides a mix of migratory fish on their way north or south, depending on the season. It is a prime fishing ground for tuna, bluefish, swordfish, marlin, dolphin, king mackerel, and sea bass, among other species. State marine officials have calculated the dockside value of the winter trawl fishery alone at $7.5 million to $10 million a season, which accounts for a quarter to a third of the marketable food fish landed in North Carolina today.

Although the exploratory drilling itself would not affect this abundant catch, marine experts worry that if gas or oil is discovered and other wells are drilled, discharges of "mud" and other toxic substances associated with drilling will have a deleterious effect on the sargassum beds that serve as habitat for marine animals that become food for predatory fish. The sargassum beds tend to fetch up along the edge of the Gulf Stream because of the convergence of different water masses there. From the fisherman's point of view, Mobil could not have chosen a more environmentally sensitive area for its drilling. While the evidence of serious negative impact from a single exploratory drilling itself is not compelling, concerns about the effect of widespread drilling and exploration in the area cannot be dismissed.

Considering the historical and environmental significance of the waters off Cape Hatteras, and the fact that Carolina's coast has never before yielded oil or gas, it is not surprising that Mobil's proposal has attracted national and worldwide attention. It symbolizes the kinds of energy choices the United States will face if we do not wean ourselves from a growing dependence on foreign sources of oil and natural gas. But where can domestic reserves be found?

The easy domestic supplies of oil and gas have already been tapped. The eastern continental shelf is one of the few remaining targets for new domestic exploration. Mobil has effectively exploited this point, noting the dependence of coastal industries such as fishing and tourism on steady supplies of fuel and the threat to foreign supplies posed by the recent Persian Gulf war. A Mobil spokesman even invoked the memory of the Wright brothers in defense of the company's plans. Without the kind of risk-taking the Wrights exhibited, there might not be modern airplanes or the fuel to propel them.

Thus Bankers can no longer count on their splendid isolation from the mainland and the protection afforded by the National Park Service to preserve their way of life. Indeed, the unfolding drama over natural gas off Cape Hatteras has become part of a larger national energy dispute. North Carolina's offshore waters are not the only target of new exploration for oil and gas. California, Florida, and Alaska are also on the list. Partly because of intense lobbying by environmental groups and by the governors and congressional delegations in several coastal states, in June 1990, President Bush announced a ten-year moratorium on offshore drilling for California, Florida, Alaska, and much of the mid-Atlantic coast from Maryland to the Connecticut–Rhode Island border. But to the astonishment and anger of North Carolina governor Jim Martin and other state officials, North Carolina was not included on that list. The president's decision may prove to be a critical turning point in the Mobil dispute, because it prompted the governor to announce unequivocal opposition to offshore drilling and assured state resistance along every step of the process.

It also mobilized local congressman Walter Jones, who attached his legislation postponing action on any offshore North Carolina exploration to a popular congressional bill governing oil spills. That tactic worked. For a moment, at least, the drive to find gas off Cape Hatteras was stalled. The legislation imposed a moratorium on any offshore activity until October 1991, and prohibited any further movement until "adequate physical oceanographic, ecological, and socioeconomic information is available to enable informed decisionmaking. . . ." But can the congressman, governor, and their Outer Banks constituents indefinitely forestall the enormous pressure to drill? The Bush administration is stongly committed to finding new domestic sources of fossil fuel, especially after the experience of the Persian Gulf war. A new national energy plan announced during that war advocated a balance between new exploration and conservation. In the view of many environmentalists, the plan fell far short of the strong conservation measures required to reduce U.S. dependence on petroleum, both foreign and domestic.

Meanwhile, Americans answered with their gas pedals, increasing U.S. dependency on foreign oil to near the 50 percent level in 1990 for the first time since the OPEC oil embargo of 1973–74.

It was not an auspicious sign for the Outer Banks. Perhaps for the first time in their long history, the islands' isolation, their treacherous offshore waters, and their affinity for violent storms would not buffer them from a new form of human exploration. In the immediate future, Americans' unquenchable thirst for gas and oil threatens to wash over the Outer Banks more powerfully than any cyclonic storm.

ON OCTOBER 26, 1990, one might have forgiven the Bankers for not thinking so. Mobil and its petroleum rigs might loom like low, dark clouds on the autumn horizon, but the storm that had swept in upon the people of the Banks the night before had effectively shut down the islands as well as any in history.

On the afternoon of the previous day, Captain William Cliett, skipper of the dredge *Northerly Island,* had decided to knock off operations. His 200-foot dredge, owned by North American Trailing Co. of Chicago, was under contract with the Army Corps of Engineers to maintain a navigation channel through the constantly shifting shoals of Oregon Inlet. The vessel was part of an annual $4.4 million Corps program to dredge 700,000 cubic yards of sand from the inlet. It was also a bit player in a much larger debate over whether Oregon Inlet should be stabilized with a pair of mile-long jetties costing more than $100 million of taxpayers' money. The *Northerly Island* was not to remain a bit player for much longer.

According to the captain's subsequent account, the weather forecast called for twenty-five- to thirty-five-knot winds that night, with accompanying choppy seas. Captain Cliett and his crew of nine did not consider the forecast particularly ominous or unusual. He chose to anchor for the night in his usual spot east of the two-and-a-half-mile Bonner Bridge, which has spanned the inlet since 1963. But the forecast proved too benign. Within hours the weather turned foul. Cliett later said seas were running three to five feet, the winds were slicing forty-five to fifty-five knots—gale-force level—with gusts well above that mark, and the current surging to at least six knots. Subsequent investigation has shown that this storm, though short-lived, was one of the most severe of the past fifty years.

When moving water strikes a solid object it is deflected: forced to change direction. That deflected water must go somewhere. Most of it is deflected laterally—to the sides—and flows around the solid obstacle. Some, a little, goes up; more goes down, following directions from gravity. As it moves down it churns the sand around the obstacle: swirls it, suspends it, levitates it into the passing current. The sand moves on downstream.

A beachgoer can watch this happen in the waves on the beach. Just stand there in the swash facing the sea. The incoming waves undermine your toes, tipping you forward. But the backwash eddies under your heels, momentarily correcting the balance. With remarkable alacrity, you will settle deeper into the sand. But the currents you are deflecting will not be content to sink you in the sand. If you can stand still a moment or two longer, you will be toppled. Then you will be rolled. You roll because you are roughly cylindrical in cross section. A sheet of plywood or a slab of bark would simply go back and forth, up and down the beach. At the beach the current goes back and forth, in and out from the sea. In the inlet it goes out.

Glance back now at the photograph of the Banks taken from the Apollo 9 spacecraft. Those plumes or fans extending into the sea at each inlet are where North America is leaving, going out. Captain Cliett's moment and the *Northerly Island*'s time had come.

Made of heavy iron, a ship's anchor makes a nearly perfect object to deflect moving water. Its flukes give it surprising lift for its weight. As surging current undermines it, the mitigated force of water to hold it down diminishes. As the anchor moves down current, toward the strictured mouth of the inlet, the swirling, sandy bottom slips away, too, deeper and deeper. An anchor works best when the angle its cable makes with the surface is most shallow. The deeper the water, the closer the floating vessel is to being directly over its anchor, and the slimmer the chance that the anchor will hold the vessel. Nothing is more terrifying than those moments when a captain first perceives that his anchor has moved. Rarely does the situation correct itself. Things usually get precipitously worse. The anchor moves deeper; the cable's angle gets steeper; the anchor moves again, farther, faster now.

At about 10:30 P.M., Captain Cliett realized the dredge's anchor was dragging. He ordered the vessel's engines at full power and directed that the ship's ballast tanks be pumped full of water to keep the barge stranded on a shallow shoal. The gale was easterly. But the current and tide were causing the shoal to erode, and the dredge continued its inexorable sideways drift toward Bonner Bridge. The Coast Guard would later rule that the captain may have been negligent because he failed to drop a second anchor during the ship's grounding. Cliett, who was later absolved of any responsibility after an inquiry, argued that forces far more powerful than any under his control had commandeered his ship.

At 12:50 A.M., Cliett notified the Coast Guard to close the bridge to traffic. Twenty minutes later, the *Northerly Island* crashed into the bridge. Four of the dredge's terrified crew members climbed up the ship's superstructure and made their way onto the bridge. They were carried to safety by Dare County sheriff's deputies just minutes before the span collapsed. The officers watched in horror and amazement as the red-and-white dredge severed the bridge and catapulted 370 feet of the two-lane highway into the inlet. A fiery explosion caused by ruptured electrical cables lit up the gale-swept sky. Astonishingly, no one on board or on the bridge was injured.

Thus did the vessel commissioned to keep the inlet open temporarily close it. But much more than the bridge was severed that night. The structure had been the only direct link connecting the roughly 5,000 residents of Hatteras Island to the amenities of Nags Head, Manteo, and the mainland beyond. Before the bridge was built, rugged Hatteras residents were accustomed to living in isolation for long periods. The only link to Nags Head in those days was a long, circuitous ferry ride across the sound side of Oregon Inlet. But the bridge's advent made the drive to Nags Head an easy one. Hatteras residents took jobs to the north. The bridge also opened Hatteras Island to tourists and developers. The bridge even served as a handy conduit for electrical and telephone lines to the island. It was, symbolically, Hatteras's umbilical cord to the outside world.

When the cord was cut that gusty night in October, it revived the worst fears of Hatteras people. Not only were they plunged into darkness and cold with the loss of electric power, their economic lifeline had been tossed over, too. Autumn is the prime time for fishermen to make their pilgrimage to the Outer Banks. But with Bonner Bridge gone, the only access to the island was by ferry from Ocracoke, or by a new, makeshift ferry service across Oregon Inlet. The new ferries, up and running within nine days of the bridge's collapse, required an hour or more to traverse the inlet. The ferries' zigzag path covered six and a half miles; they frequently ran aground in the inlet's shifting shoals. Whereas the bridge could accommodate a continuous stream of traffic amounting to an average of four to eight thousand a day, a single ferry could carry a scant thirty-four cars each trip.

Wholly dependent on tourism, the island's economy plummeted. Attendance at the Cape Hatteras Lighthouse in November dropped by some 7,000 visits from the previous November. Motels, restaurants, and charter boats stood empty. As if the islanders had not been punished enough, two weeks after the bridge's collapse a tornado skipped across the island. The capricious twister's seventy-five-mile-an-hour winds destroyed three new homes and about a dozen mobile homes, and ripped part of the roof off Bubba's Bar-B-Que in Avon. Cinder blocks lifted from the restaurant's walls crushed a car in the parking lot.

Well-schooled in the art of survival, the Bankers shrugged their shoulders and waited. A few old-timers welcomed the serendipitous return to the slower pace of prebridge days. Initial predictions were that repair of the bridge would consume at least six months. But an unusually mild winter and round-the-clock repair work brought the structure back into service by mid-February 1991—less than four months after its collapse.

Apart from its immediate impact on Hatteras Island, the bridge's collapse enflamed an old debate about the Oregon Inlet jetties and may have hastened its resolution—much to the consternation of geologists and conservationists. Twenty years before, Congress had authorized construction of mile-long jetties at either end of Oregon Inlet. The jetties were intended to stabilize the inlet and permit easier passage for commercial and recreational boats. But like a lot of grandiose schemes on the Outer Banks, this one had gone nowhere—until the bridge's collapse distilled the political glue that just may stick the jetty plan together.

Among the most enthusiastic backers of the jetties is a handful of commercial fishermen based in Wanchese, a fishing community on Roanoke Island. In 1976, the state of North Carolina began constructing a seafood industrial park in Wanchese. The park was intended to serve as a major seafood processing center for trawlers working off the Outer Banks, and thus as a generator of jobs and economic growth in the region. The park's economic viability was predicated on construction of the jetties, which the fishermen said would stabilize the inlet and permit safe passage between the shoals and under Bonner Bridge. But the jetties' high price tag—more than $100 million at that time—and persistent environmental concerns stymied the project.

Absent the jetties, the seafood park lost money every year. Seafood processors deemed it too risky an investment because of the inlet's unpredictable shoals. On many days of the year, fishing boats were simply unable to navigate the inlet's shallow waters. Some of the fishing trawlers wrecked and ran aground. Alarmed, the owners of these modern, high-tech vessels hauled their catch to seafood processors in Cape May, New Jersey, and New Bedford, Massachusetts. By mid-1991, only two processing plants were operating in the Wanchese park.

Wanchese fishermen, backed by the Army Corps of Engineers, seized on the bridge's collapse as evidence for their case. Had the jetties been in place, there would have been no need for the dredge, and thus no opportunity for the accident. But the bridge's temporary demise also invites a different interpretation: that man's attempts to keep the inlet open—whether by dredging, jetties, or other means—are ultimately futile.

That Oregon Inlet is shifting is not in dispute. Like the birds that nest there, the inlet is migrating south—and at a rate of 100 feet per year. Opened by a fierce hurricane on September 7, 1846, Oregon Inlet has behaved as inlets are supposed to: longshore currents are shifting its location from north to south, while a combination of currents, wind, and wave action is filling the inlet in. The net transport of sand south from the inlet has been estimated at 710,000 cubic meters per year. The inlet has also been moving landward at a rate of five meters per year. The effect of these factors is that the northern tip of Pea Island is disappearing and exposing the southern end of Bonner Bridge to severe wave action.

Even as the debate over the jetties raged during 1990–91, state officials constructed a smaller, 2,500-foot stone revetment and groin at the inlet's south end at a cost of $15 million. Much of this structure caps the present north end of Pea Island, but some 800 feet extend out to sea. Its purpose is to protect the bridge from powerful waves and to prevent the land to which it is attached from washing away. The north end of Pea Island had been eroding at a rate of 180 feet a year. A single, ferocious northeaster in March 1989, ripped away 200 feet of the island's north tip in one night. The same storm destroyed beach houses in Nags Head and Kitty Hawk and caused more than $4.5 million in damage up and down the Banks. But that is a rather ordinary storm for the neighborhood. The Pea Island project had long been blocked by the Interior Department, which manages the Pea Island National Wildlife Refuge on which the groin was to be built. But strong state political pressure and the realization that the highway might be separated from the bridge presumably persuaded officials to relent and grant a permit for the groin.

The argument against groins and jetties is that, far from preventing erosion, such man-made devices contribute to erosion by blocking sand transport and starving beaches downstream. In the case of the Oregon Inlet jetties, geologists such as Stan Riggs think the consequences of the jetties are obvious: they will deny the natural flow of sand to Pea and Hatteras islands and accelerate the erosion that is already carving into both islands. At the same time, the historical tendency of the Banks to move landward, and of Oregon Inlet to shift dramatically south and west, makes it an ultimate certainty that the jetties will not perform their intended function for long. The pace of the inlet's abandonment of Bonner Bridge may be slowed and slightly altered by the jetties, but it will not be stopped—especially if a violent storm on the magnitude of the 1846 hurricane should sweep through. "The dynamics of that system are such that you cannot build a structure like that without some consequences," says Riggs. "It is going to change the [beach] profile, it is going to change the sediment flow, and anybody who thinks any differently has got his head in the sand."

Protecting the bridge is only the subplot in an even larger drama, just as the inevitable tendency of Oregon Inlet to close is part of the larger barrier island dynamic. As the inlet shoals up, pressure builds elsewhere for a new inlet to open. Several miles south of Oregon Inlet, between Pea Island and Chicamacomico Banks, just north of the villages of Salvo and Waves, is the site of a former inlet, "New Inlet," which remained open until 1922. Its ancestor was old Chickinacommock. It made Pea Island an island. This inlet reopened for a few weeks after the Ash Wednesday Storm of March 1962. Indeed, in the centuries before Oregon Inlet opened, at least a dozen inlets opened and closed between False Cape, Virginia, and Cape Hatteras. There is little reason to suppose that, unlike every one of its ancestors, Oregon Inlet can be frozen in its tracks and prevented from closing, or another "new" inlet from opening.

Bonner Bridge is not the only man-made structure jeopardized by this natural process. The highway that winds south from it to Hatteras is threatened, too. Now that the islands' dune systems are no longer stabilized, the highway is increasingly exposed to overwash from the ocean side. Whether or not a new inlet opens again south of Oregon Inlet, storm surges will inevitably spill over the highway at other low, narrow points along Hatteras Island. Even if Bonner Bridge is protected, the possibility of frequent highway washouts is real. If you understand sand transport (explained in Chapter Three) you understand why that risk will grow if the jetties are built and erosion to their south accelerates.

Opposition from geologists, environmentalists, and officials of the Cape Hatteras National Seashore has not deterred advocates of the jetties. On the eve of his hard-fought reelection in November 1990, Senator Jesse Helms announced that the jetties would be built after all. The Interior Department withdrew its opposition to the jetties in exchange for having a voice in the system's design. Helms cited the accident at Bonner Bridge as the impetus for the sudden evaporation of top Interior Department opposition to the jetties. The fact that Helms was joined in his lobbying by senior elected officials from across the state contributed to the department's apparent change of heart.

But Helms's announcement did not guarantee that the

jetties will be built. The Interior Department and the Corps of Engineers have commissioned a team of scientists, including Robert Dolan, to review all the issues surrounding the Oregon Inlet controversy. That review could still sway any final decision. The project must also win budget approval from the White House and from Congress in a time of federal cutbacks.

There is one way, of course, to predict accurately the impact of a jetty system on the beaches to the south of Oregon Inlet. That is to study the effect of the new, smaller rock groin at the south end of Bonner Bridge on the transport of sand from north to south. If beach erosion on Pea and Hatteras islands is demonstrably accelerated by this structure, then it is logical to conclude that the effect of the Oregon Inlet jetties, which would extend more than twice as far out to sea than the Bonner Bridge groin, would be substantially greater. Just such a study is part of the wider review that Dolan and his associates are conducting. This is no longer the inexact science Dolan encountered when he first came to the Banks thirty years ago. Modern instruments allow him to examine small segments of the beach and to predict, with nearly pinpoint accuracy, what their profiles will be.

The uncertainty is not in the science, but in the politics of the Oregon Inlet dispute. If Dolan and other scientists can prove that the jetties would have an adverse impact on the beaches of the Cape Hatteras National Seashore, will the government decide to abandon the project? Or will the perceived short-term economic benefits associated with the jetties override long-term environmental concerns that are not just environmental or aesthetic but ultimately financial as well?

≈

FROM THE SPAN of Bonner Bridge today one can see the processes of sand transport far better than in any textbook. To the present traveler, the bridge seems to begin well inland on the northern, upstream side of Oregon Inlet. Vast shoals showing every stage in ecological succession from bare sand, through submarine eel grass, to *Spartina* marshland, and even incoming shrubs, are decorated in season by a changing myriad of birds—including pelicans. Before the advent of hard pesticides, pelicans nested as far north as Ocracoke. Then they perished. DDT and its allies did to pelicans what they never managed to do to mosquitos: extirpate them. But, DDT banned, pelicans have come back. And not just back. As the climate has warmed in the past few decades, pelicans have been among the many animal and bird species of the Deep South to expand their ranges northward. Good for the pelicans; they may actually be benefiting from the greenhouse effect.

As one crosses the span, the stricture of the channel and its displacement southward are strikingly apparent. Coming down on the Pea Island side, take a few moments to investigate the southern footings of the bridge and that little groin and revetment. As the pelicans have moved north the inlet has moved south. The pelicans have returned, but the sand has obviously gone away. The sand went south, and not just for the winter.

We believe any fair-minded person can see that Oregon Inlet wants to close. Pumping the channel prevents that, prevents a lot of sand from moving across to replenish the sand that went south. The outwash current and the longshore current bring constant streams of water against the solid barriers of those revetment rocks. The water is deflected, the sand is levitated, suspended, swept on downstream. Channel dredging and the present rock structures are denying Pea Island its sand budget. One can quickly perceive what will happen when the *north* side—the Bodie Island side, the growing side—of the inlet is capped with a rock wall and jettied out to sea. Longshore sand transport will obviously be far more dramatically curtailed than it is even now. That is, after all, the precise purpose of the proposed jetties—to "stabilize" the inlet.

And to the south, downstream? Fewer than thirty miles downstream stands a temporal counterpoint to the new problems centered on Oregon Inlet: an old problem with a most immediate, connected future. The problem is succinctly and accurately described in a 1988 report by a distinguished panel of the National Research Council, an arm of the National Academy of Sciences: "Cape Hatteras Lighthouse, the tallest and best-known brick lighthouse in the U.S., faces destruction due to coastal erosion."

The etiology of the deteriorating condition of the Cape Hatteras Lighthouse dates back to the origin of the Outer Banks as we know them today. This time the controversy is more symbolic, financial, and historical than environmental, although sand and sea assume a pivotal role. The Council's term, "coastal erosion," may be something of an understatement. As its report makes clear, a single 100-year storm could topple the 200-foot lighthouse at any time. Such a storm, which is defined as having a 1 percent probability of occurring in any year, is estimated to have a surge potential of 8.8 feet above sea level. In its current vulnerable position, the lighthouse is only 8 feet above sea level, which means that waves in a storm surge would engulf its base. Worse, scientists calculate that the average height of highest waves breaking *above the storm surge elevation* would be 15.5 feet. Such a storm, in the report's words, "would directly attack the lighthouse, undermine its shallow footings, and probably demolish its accessory buildings as well.'' Deflected water again. Thus might a national historic landmark, symbol of the Atlantic coastline and guardian of the Graveyard of the Atlantic, topple spectacularly into the sea.

When the current lighthouse was built in 1870—the second to stand on that site—the shoreline was 1,500 feet away. Today it is about 200 feet away. There is no more visible or convincing evidence of sea-level rise and beach migration anywhere on the Outer Banks. In the recent past the sea has encroached to within 160 feet. In 1989–90, however, there was some accretion of sand at Hatteras. The mild 1990–91 winter did not produce the sort of severe storm that characteristically reclaims such accretion and takes more ground as interest on the loan. The respite was only temporary, as the fierce ''Halloween'' storm of late October 1991 demonstrated. The storm—which struck coastlines from Maine to Florida—wiped out dunes, overwashed the highway on Hatteras Island, and further eroded sand at the base of the lighthouse.

The Research Council report documents an average rate of shoreline retreat for the vicinity of the lighthouse at about twenty feet a year. That retreat is not steady; there are ups and downs. But add to that calculus a sea-level rise of .05 inches a year, and the report estimates that the sea level at Cape Hatteras will go up nearly 8 inches in 100 years. Those figures assume that the trends of the past century will remain unchanged. If sea-level rise accelerates as a consequence of global warming, however, at Cape Hatteras thirty years from now it could be 6.1 inches higher, and 50 inches higher in a century. And those are conservative projections.

If current sea-level trends continue, the shoreline is predicted to retreat 157 feet landward by the year 2000; using the highest sea-level estimates, that retreat would be 407 feet. In either case, the lighthouse will be washed away. No wonder the National Park Service has selected the Outer Banks as one of several sites at which to analyze the impact of global warming on barrier islands. Where else would sea-level rise be more visible and measurable? Measurement or not, the reality of global warming is not in doubt: 1990 was the warmest year of any since records have been kept, and six of the warmest years since 1850 have occurred during the past decade.

The more immediate question is what to do about the lighthouse. Apart from its historical significance, the lighthouse is a major tourist attraction and a part of the Cape Hatteras National Seashore. Trapped in the middle as usual, the National Park Service is directed not to use artificial means to preserve man-made structures on the one hand, but is obligated by law to preserve historic landmarks on the other. The contradiction in this case has clearly been resolved in favor of the lighthouse.

Efforts to keep the sea from claiming the lighthouse date to 1930, long before the national park existed. Groins, protective dunes, revetments, beach nourishment, and even artificial seaweed have all been deployed to slow the ocean's advance. After several years of study, the Army Corps of Engineers proposed in 1985 that the lighthouse be protected by an octagonal concrete seawall at its base. The top of the seawall would rise 23 feet above sea level, and its outward face would be protected by a stone revetment that would extend more than 200 feet into the sea. Only the six seaward sides of the wall would be built initially; but as the sea advanced, the wall would be completed on the landward side and the lighthouse would slowly become an island unto itself, separated from the land behind.

This proposal would violate state and federal rules against

using seawalls to protect coastal structures and would accelerate beach erosion. It would hide the base of the lighthouse and utterly alter its relationship to the sand and sea. A citizens' group persuaded the Park Service not to fortify the lighthouse, but to move it. The suggestion that the lighthouse be moved had been made before, but was rejected for fear the structure would crumble under the stress of being transported. But an extensive structural study in 1986 indicated that the lighthouse is structurally sound and could withstand a move, carefully done.

But how would such a move be accomplished? The committee proposed that the lighthouse be structurally repaired, then underpinned by a series of steel beams that would be lifted by hydraulic jacks. Once lifted, the lighthouse would be lowered onto rollers resting on horizontal steel-rail beams supported by concrete piles. The lighthouse would then be rolled, very slowly, to a new concrete mat placed behind its current location. Plans to move it in stages of a few hundred feet at a time have eventually given way to a grander scheme: move it 2,500 feet southwest so that it will come to rest 1,500 feet inland—its original position relative to the ocean due east. The estimated total cost for moving the lighthouse and associated structures was $9 million in May 1991.

Is there precedent for such a move? Yes: everything from a 12,000-ton, fourteenth-century church in Czechoslovakia (moved 2,400 feet in 1975) to oil rigs as tall as 200 feet and weighing up to 35,000 tons have been moved onshore and loaded onto barges without mishap. The Hatteras Lighthouse, after all, weighs only 2,800 tons. But then the church in Czechoslovakia and the oil rigs probably did not have to contend with a major potential threat to the lighthouse move—the chance that a hurricane might blow through as the structure tottered on its rails. The obvious answer is not to move it during hurricane season, but at any season the risk of a storm is a regular part of life at Cape Hatteras.

Another major consideration in moving the lighthouse is how the change would affect Buxton Woods, its new location. The move would require some destruction of trees on the edge of the Woods, at least. Although the heart of the maritime forest would remain protected, an increase in activity around the lighthouse would represent a potential threat. The buildings associated with the lighthouse would also be moved, and a new parking lot built. A considerable amount of the rich edge communities called pine thicket and shrub savannah would be lost.

The Park Service has decided to bide its time until the lighthouse can be repaired, funds are appropriated, and erosion poses a more immediate threat. Meanwhile, there is one option that the committee did not consider, because it was not part of the committee's official charge. That is, to leave the lighthouse alone and let the sea claim it. After all, the lighthouse long ago relinquished its function as a navigational aid; to move it farther back would transform it into a museum and completely obviate the purpose for which it was intended. But even Duke University geologist Orrin Pilkey, Jr., a strong advocate of letting the sea have its way, agrees that the lighthouse is too closely linked to the Outer Banks and the nation to be destroyed by neglect. He approves of the move because it avoids the use of a seawall and sets a good precedent for the disposition of future structures threatened by erosion. As Robert Dolan and Paul Godfrey argued years ago, a few structures on the Outer Banks are fine, as long as they are mobile and can be shifted to accommodate the rising sea and its attendant overwash.

We leave the lighthouse controversy where we found it:

unfinished, still raging, confused. In early 1991, the National Park Service began a million-dollar renovation. Expected to take thirteen months, the repairs will fix the roof, reinforce mortar in the interior and exterior walls, and make the interior stairs, windows, and railings safe again. Once completed, the refurbishment should make it possible for visitors once again to climb to the top of the lighthouse and peer out over Diamond Shoals, Buxton Woods, and Pamlico Sound. The Park Service plan is sensible. To be moved safely, the structure must first be fortified. Additional funds must also be appropriated by Congress to finance the move. Further studies must be done, and permits secured. There is time, though not much. Nature's patience is limited. If the stately lighthouse is to be preserved for future generations, it, too, must beat a hasty retreat before the sea's inevitable march.

We view the decision to move the lighthouse as a victory for good sense and the preservation of something genuinely valuable. But we see it in a much broader context, and that context is alarming. The delicately balanced trilogy of rice rat, kingsnake, and wax myrtle explored in Chapter Six mirrors the resiliency and self-correcting mechanism of the natural order. But humans have not been so careful or clever in their stewardship of the earth. Compare the trilogy examined earlier to the wildly out-of-balance trio of global warming, petrochemicals, and overpopulation.

The sea-level rise that threatens the Cape Hatteras Lighthouse is now unnaturally rapid as a result of the greenhouse effect and global warming. Indeed, sea level might have naturally stopped rising sometime before A.D. 1800 but for the Industrial Revolution. Global warming proceeds from combustion of fossil fuels and concomitant deforestation. The vast tonnage of CO_2 produced is not consumed by earth's plants because every minute there are fewer of them. And the petrochemical industry does not produce just greenhouse gases, or just toxic wastes. It produces commercial toxins such as the DDT that decimated the pelicans and still poisons the life support systems of much of the Third World. These excesses cut ever more deeply into the ecosystemic roots that sustain us.

Pesticides, oil spills, and clearcutting are but manifestations of an even more fundamental imbalance: our egregious overpopulation. In the rising tide of our numbers it is cruel folly to talk of significantly raising the standard of living for any large number of people. To raise the standard of living for the people of India, for example, to that of the average American, would—using any technology now on the drawing boards—decimate all life on earth in a stifling blanket of heat and pollution, exhausting any supply system imaginable. Current discussion of human overpopulation in the Western world is just cocktail party repartee between the rich and the extremely wealthy. With callous insensitivity, our elected leaders deny the technology and education of population control to the mass of humanity, at home and abroad. Anyone can see the fundamental, undeniable connection between all our problems and the one viable solution.

The news is not all bleak. The lighthouse can be saved, despite revetments and sea-level rise. And the pelicans have increased, despite oil spills and the *daily* loss of 75,000 acres of earth's tropical forests. Synthetics from petrochemicals make wonders like nylon and detergents that brighten our lives, despite the fact that two-thirds of the world's children are so malnourished they will never develop their mental potential, and six tons per acre of North Carolina's cropland topsoils are eroding away every year.

These overarching realities make our encounters with nature's balance and order all the more precious. The respites they provide seem somehow shorter, softer, more aesthetic and less pragmatic than those strident problems. We must quickly heed the pragmatic, of course. But we can linger upon and cherish the aesthetic, however precarious and fleeting our suspension of this grim trend. The Outer Banks envelop both.

CHRISTMAS ON THE Outer Banks is a time that belongs to the islanders themselves, that reunites families and friends. No holiday there was more magical than the Christmas of 1989. Overall, 1989 was a warm year, one of the majority this decade that signal the greenhouse effect and climatic change. But for a few brief days at the year's end, the beaches of Hatteras were not only white with sand. They were gleaming under cover of a heavy snowfall that reached far into the Deep South.

It was Friday, December 23, and Ocracoker Rudy Austin was getting ready for work. As ferryboat captain on the Ocracoke to Hatteras run, he handled the night shift. It was then that he noticed it, noticed the big flakes of snow descending from a slating sky. The flakes blanketed the picturesque village—the docks, the lighthouse, the big boat in Rudy Austin's front yard. The snow fell aimlessly on the marshes and beaches, too, creating a scene as incongruous as a seascape in Kansas. The snow swirled around the Cape Hatteras Lighthouse and decorated the Lost Colony grounds on Roanoke Island. All along the coast of North Carolina, from Manteo to Wilmington, it was snowing with abandon.

Though rare, snow is not unheard of on the Carolina beaches. It usually comes later in the winter and does not linger. But this time it snowed as if it would never stop. It fell for two straight days; the accompanying wind and precipitous cold mounded it into big drifts. Roads were impassable.

The snowfall dispelled a number of coastal myths, among them that it cannot snow if it is too cold—thermometers read fifteen degrees that day—that it cannot snow at the beach, and that "four-by-fours," vehicles that are the official badges of the coastal esoteric, can maneuver through anything. On this Christmas, they could not. It took one couple four hours to drive a mile and a half from their home to rescue Christmas presents stored at the office.

But nowhere was the snow more miraculous than on the Outer Banks. On Ocracoke, year-round residents like the Austin family celebrated their first white Christmas in memory. By the time the snow stopped falling, fifteen inches had been measured, a record for any winter in any year. There were coatings of snow on the dunes, like toppings on an ice cream cone. All of the Outer Banks, their houses and hammocks, their marinas and marshes, stood white-frocked.

On Ocracoke, a brisk northwest wind drifted the snow, shoveling it across the dunes and down the beach. Only the ocean kept its appointed routines. The snowflakes descended on the water, too, and, like all precipitation, tended to calm the waves and dampen their velocity. But the highest wave of the high tide on Christmas Day at Ocracoke greeted the moving snowdrift and captured it. Succeeding waves washed away snow from the beaches, scouring the sand with their cleansing power. At sea, falling flakes dissolved and were dispatched on the long odyssey whence they came, and will come again.

References

Following is a list of selected works cited in the text or used in research for these chapters.

Chapter One: *Sand*

Burk, C. A., and C. L. Drake, eds. *The Geology of the Continental Margins*. New York: Springer-Verlag, 1974.

DeBlieu, Jan. *Hatteras Journal*. Golden, Colo.: Fulcrum, 1987.

Fairbridge, R. W. *The Encyclopedia of Geomorphology*. New York: Reinhold Book Corp., 1968.

Flint, R. F. *Glacial and Quaternary Geology*. New York: John Wiley & Sons, 1971.

Holmes, A., and D. L. Holmes. *Principles of Physical Geology*. 3rd ed. New York: John Wiley & Sons, 1978.

Lazell, James. "Marine turtles in India." *Copeia* 2(1980): 374–75.

Martof, Bernard S., et al. *Amphibians and Reptiles of the Carolinas and Virginia*. Chapel Hill: University of North Carolina Press, 1980.

Rukeyser, Muriel. *The Outer Banks*. 2nd ed. Greensboro: Unicorn Press, 1980.

Snead, R. E. *Coastal Landforms and Surface Features: A Photographic Atlas and Glossary*. Stroudsburg, Pa.: Hutchinson Ross, 1982.

Webster, William David, et al. *Mammals of the Carolinas, Virginia, and Maryland*. Chapel Hill: University of North Carolina Press, 1985.

Chapter Two: *Land*

Coates, Donald R., ed. *Coastal Geomorphology*. Binghamton: State University of New York, 1973.

Colquhoun, D. J. "Holocene sea-level curve for South Carolina, U.S.A. fluctuation implication." *Late Quaternary Sea-Level Changes*. Edited by Qin Y. and Zhao S. Beijing: China Ocean Press, 1987.

Emery, K. O. "The continental shelves." *Scientific American* 221(1969): 106–22.

Gornitz, V., S. Lebedeff, and J. Hansen. "Global sea level trend in the past century." *Science* 215 (1982): 1611–14.

Johnson, Douglas Wilson. *Shore Processes and Shoreline Development*. New York: John Wiley & Sons, 1919.

Kerr, Richard A. "Climate since the ice began to melt." *Science* 226 (1984): 326–27.

Lazell, James. "The Outer Banks of Brazil." *The Explorers Journal* 68 (1990): 97–113.

Leatherman, Stephen P., ed. *Barrier Islands: From the Gulf of St. Lawrence to the Gulf of Mexico*. New York: Academic Press, 1979.

MacLeish, William H. *The Gulf Stream: Encounters with the Blue God*. Boston: Houghton-Mifflin Co., 1989.

Milliman, J. D., and K. O. Emery. "Sea levels during the past 35,000 years." *Science* 162 (1968): 1121–23.

Pethick, J. *An Introduction to Coastal Geomorphology*. London: Edward Arnold, 1984.

Stahle, D. W., M. K. Cleaveland, and J. G. Hehr. "North Carolina climate changes reconstructed from tree rings: A.D. 372 to 1985." *Science* 240 (1988): 1517–19.

Stick, David. *Roanoke Island: The Beginnings of English America*. Chapel Hill: University of North Carolina Press, 1983.

Titus, Jim. "Sea level rise and barrier islands." *Bulletin of the Coastal Society* 7 (1983): 4–7.

Trefil, James. "Modeling Earth's future climate requires both science and guesswork." *Smithsonian*, December 1990, 29–37.

Chapter Three: *Water*

Bailey, Anthony. *The Outer Banks*. New York: Farrar, Straus and Giroux, 1989.

Ballance, Alton. *Ocracokers*. Chapel Hill: University of North Carolina Press, 1989.

Behn, Robert D., and Martha A. Clark. "Termination II: How
the National Park Service Annulled Its 'Commitment' to a
Beach Erosion Control Policy at the Cape Hatteras National
Seashore." Working Papers, Institute of Policy Sciences.
Durham: Duke University, 1976.

Byrd, William. *Histories of the Dividing Line betwixt Virginia and
North Carolina*. Edited by William K. Boyd. Raleigh: North
Carolina Historical Commission, 1929.

Cumming, William P. *Mapping the North Carolina Coast: Sixteenth-
Century Cartography and the Roanoke Voyages*. Raleigh: North
Carolina Department of Cultural Resources, Division of
Archives and History, 1988.

Dolan, Robert. "The Ash Wednesday Storm of 1962: 25 Years
Later." *Journal of Coastal Research* 3 (1987): ii–vi.

———. "Barrier Dune System along the Outer Banks of North
Carolina: A Reappraisal." *Science* 176 (1972): 286–88.

———, Paul J. Godfrey, and William E. Odum. "Man's Impact
on the Barrier Islands of North Carolina." *American Scientist* 61
(1973): 152–62.

———, and Bruce Hayden. "Adjusting to Nature in Our National
Seashores." *National Parks & Conservation Magazine*, June
1974, 9–14.

———, Bruce P. Hayden, and Kenton Bosserman. "Is the Lost
Colony Really Lost?" *Shore and Beach*, April 1981, 17–19.

Dunbar, Gary S. *Geographical History of the Carolina Banks*. Technical
Report No. 8. Baton Rouge: Louisiana State University, 1956.

Godfrey, Paul J. "Barrier beaches of the East Coast." *Oceanus* 19
(1976): 27–40.

———. *Oceanic Overwash and Its Ecological Implications on the Outer
Banks of North Carolina*. Washington: National Park Service,
1970.

———, and Melinda M. Godfrey. *Barrier Island Ecology of Cape
Lookout National Seashore and Vicinity*. Monograph Series 9.
Washington: National Park Service, 1976.

———. "Comparison of ecological and geomorphic interactions
between altered and unaltered barrier island systems in North
Carolina." Edited by D. R. Coates. *Coastal Geomorphology*.
Binghamton: State University of New York, 1973.

———. "The role of overwash and inlet dynamics in the formation of salt marshes on North Carolina barrier islands." *Ecology
of Halophytes*. New York: Academic Press, 1974.

Hartzer, Ronald B. *To Great and Useful Purpose: A History of the
Wilmington District U.S. Army Corps of Engineers*. Wilmington,
N.C.: U.S. Army Corps of Engineers, 1984.

Herndon, G. Melvin. "The 1806 Survey of the North Carolina
Coast, Cape Hatteras to Cape Fear." *North Carolina Historical
Review* 49 (1972): 242–53.

Kaufman, Wallace, and Orrin H. Pilkey, Jr. *The Beaches Are Moving:
The Drowning of America's Shoreline*. Durham: Duke University
Press, 1983.

Leatherman, Stephen P., ed. *Barrier Islands: From the Gulf of St.
Lawrence to the Gulf of Mexico*. New York: Academic Press,
1979.

———. *Barrier Island Handbook*. Amherst, Mass.: National Park
Service Cooperative Unit, 1979.

———, ed. *Barrier Islands*. New York: Academic Press, 1979.

Ludlum, David M. *Early American Hurricanes, 1492–1870*. Boston:
American Meteorological Society, 1963.

Schwartz, Maurice L. *Barrier Islands*. Benchmark Papers in Geology. Stroudsburg, Pa.: Dowden, Hutinson & Ross, 1973.

Stick, David. *Graveyard of the Atlantic: Shipwrecks of the North Carolina Coast*. Chapel Hill: University of North Carolina Press,
1952.

———. *The Outer Banks of North Carolina, 1584–1958*. Chapel
Hill: University of North Carolina Press, 1958.

Chapter Four: *Blackbeard*

Cumming, William P. "The turbulent life of Captain James Wimble." *North Carolina Historical Review* 46 (1969): 1–8.

———. "Wimble's maps and colonial cartography of the North
Carolina coast." *North Carolina Historical Review* 46 (1969):
157–70.

Epperson, D., et al. *Weather and Climate in North Carolina*.
Raleigh: North Carolina Agricultural Research Service, 1988.

Gosse, Philip. *The History of Piracy*. New York: Tudor, 1934.

Johnson, Charles. *A General History of the Pirates*. 1724. Cayme
Press edition. Kensington, England: Stanhope Mews West,
1925.

Lee, Robert E. *Blackbeard the Pirate*. Winston-Salem: John F. Blair, 1974.

Powell, William S. *North Carolina Through Four Centuries*. Chapel Hill: University of North Carolina Press, 1989.

Pyle, Howard. *Book of Pirates*. New York: Harper & Brothers, 1921.

Sherry, Frank. *Raiders and Rebels: The Golden Age of Piracy*. New York: Hearst Marine Books, 1986.

Snow, Edward Rowe. *Pirates and Buccaneers of the Atlantic Coast*. Boston: Yankee, 1944.

U.S. Department of Commerce. *Climatic Atlas of the United States*. NOAA reprint. Asheville, N.C.: National Climatic Center, 1977.

Chapter Five: *Woods*

Au Shu-fun. *Vegetation and Ecological Processes on Shackleford Bank, North Carolina*. Washington: National Park Service, 1974.

Braswell, Alvin L. "Amphibians and reptiles of Buxton Woods." *Proceedings of the 1989 Symposium on Barrier Islands*. National Park Service, in press.

Bratton, Susan P., and Kathryn Davison. *The Disturbance History of Buxton Woods, Cape Hatteras, North Carolina*. CPSU Technical Report 16, Cooperative Unit, National Park Service. Athens: University of Georgia Institute of Ecology, 1985.

Brown, C. A. *Vegetation of the Outer Banks of North Carolina*. Baton Rouge: Louisiana State University Press, 1959.

———, C. J. Burk, and H. A. Curran. "Palynological analyses of a dated peat core from the North Carolina Outer Banks." *ASB Bulletin* 31 (1984): 51–52.

Burk, C. John. "The North Carolina Outer Banks: a floristic interpretation." *Journal of the Elisha Mitchell Scientific Society* 78 (1962): 21–28.

———, H. A. Curran, and T. M. Czerniak. "Dunes and vegetation: natural recovery on a damaged barrier island." *Shore and Beach* 49 (1981): 21–25.

Burney, D. A., and L. P. Burney. "A paleo-ecological investigation of Nags Head Woods Ecological Preserve, Dare County, North Carolina." Unpublished report to the Nature Conservancy, Arlington, Va., 1984.

Cairns, Huntington. *This Other Eden*. Kitty Hawk, N.C.: School of Applied Ontology, 1973.

Conant, R., and J. Lazell. "The Carolina salt marsh snake: a distinct form of *Natrix sipedon.*" *Breviora*, Museum of Comparative Zoology 400 (1973): 1–13.

DeBlieu, Jan. "Saving Buxton Woods." *CAMA Quarterly*. Fall 1988, 6–15.

Dunbar, G. S. *Historical Geography of the North Carolina Outer Banks*. Baton Rouge: Louisiana State University Press, 1958.

Jones, Lu Anne, and Amy Glass. "Everyone Helped His Neighbor." *Memories of Nags Head Woods*. Kill Devil Hills, N.C.: The Nature Conservancy, 1987.

Lewis, I. F. *The Vegetation of Shackleford Bank*. Economic Paper 46. Raleigh: North Carolina Geological and Economic Survey, 1917.

Lopazanski, M. J., J. P. Evans, and R. E. Shaw. *An Assessment of the Maritime Forest Resources on the North Carolina Coast*. Raleigh: North Carolina Department of Natural Resources and Community Development, 1988.

Ogburn, Charlton, Jr. *The Winter Beach*. New York: William Morrow, 1966.

Oosting, H. J. "Ecological processes and vegetation of the maritime strand in the southeastern United States." *Botanical Review* 20 (1954): 226–62.

———, and P. E. Bourdeau. "The maritime live oak forest in North Carolina." *Ecology* 40 (1959): 148–52.

Radford, A. E., H. E. Ahles, and C. R. Bell. *Manual of the Vascular Flora of the Carolinas*. Chapel Hill: University of North Carolina Press, 1968.

Shaw, Rich. "Ensuring a Future for Buxton Woods." *Footnotes*, newsletter of the North Carolina Sierra Club, August-September 1988, 1.

Spitzer, N. C. "Mammals." Edited by N. J. Reigle. *A Preliminary Resource Inventory of the Vertebrates and Vascular Plants of Cape Lookout National Seashore*. Management Report 22. Beaufort, N.C.: Cape Lookout National Seashore, 1977.

Webster, W. D., and C. L. Reese. "Patterns of mammalian zoogeography on North Carolina's barrier islands, with special reference to Cape Hatteras National Seashore." *Proceedings of the 1989 Barrier Island Symposium*. National Park Service, in press.

Chapter Six: *Trilogy*

Barbour, Thomas, and William Engels. "Two interesting new snakes." *Proceedings of the New England Zoological Club* 20 (1942): 101–4.

Darwin, C. *The Origin of Species by Means of Natural Selection.* 6th ed. London: Murray, 1873.

Goodyear, N. C., J. Lazell, and J. R. Alexander. "Rice rat, wax myrtle, and kingsnake: coevolution in the Intercapes Zone." *Proceedings of the 1989 Symposium on Barrier Islands.* National Park Service, in press.

Lazell, J. "Deployment, dispersal, and adaptive strategies of land vertebrates on Atlantic and Gulf barrier islands." *Proceedings of the First Conference on Scientific Research in the National Parks* 1 (1979): 415–19.

———, and J. A. Musick. "Status of the Outer Banks kingsnake, *Lampropeltis getulus sticticeps.*" *Herpetological Review* 12 (1981): 7.

Spitzer, N. C. "Rice rat's world." *Man and Nature.* Massachusetts Audubon Society, March 1973, 24–26.

Chapter Seven: *Flight*

Burlingame, R. *General Billy Mitchell.* New York: McGraw-Hill, 1952.

Combs, Harry (with Martin Caidin). *Kill Devil Hill: Discovering the Secret of the Wright Brothers.* Boston: Houghton-Mifflin, 1989.

Crouch, Tom D. *The Bishop's Boys: A Life of Wilbur and Orville Wright.* New York: W. W. Norton, 1989.

———. "In a Flight of Words: Robert Frost's Outer Banks." *Outer Banks Magazine,* 1991–92 ed., 28.

Davis, Burke. *The Billy Mitchell Affair.* New York: Random House, 1967.

DeBlieu, Jan. "Wind: Into the Dragon's Mouth." *Outer Banks Magazine,* 1990–91 ed., 18.

DeHarpporte, Dean. *South and Southeast Wind Atlas.* New York: Van Nostrand Reinhold Co., 1984.

Frost, Robert. *In the Clearing.* New York: Holt, Rinehart and Winston, 1962.

Gavreau, E., and L. Cohen. *Billy Mitchell.* New York: E. P. Dutton, 1942.

Gotch, A. F. *Birds—Their Latin Names Explained.* Dorset, England: Blandford, Poole, 1981.

Howard, Fred. *Wilbur and Orville: A Biography of the Wright Brothers*. New York: Alfred A. Knopf, 1987.

Hurley, A. F. *Billy Mitchell*. New edition. Bloomington: Indiana University Press, 1975.

Kelly, Fred, ed. *Miracle at Kitty Hawk: The Letters of Wilbur and Orville Wright*. New York: Farrar, Straus and Young, 1951.

Levine, I. D. *Mitchell*. Rev. ed. New York: Duell, Sloan, and Pierce, 1958.

Martins, Louis. *Robert Frost: Life and Talks-Walking*. Norman: University of Oklahoma Press, 1965.

Maurer & Maurer. *Aviation in the U.S. Army, 1919–1939*. Washington: USAF Office of Air Force History, 1987.

McFarland, Marvin W., ed. *The Papers of Wilbur and Orville Wright*. 2 vols. New York: McGraw-Hill, 1953.

Pearson, T. G., ed. *Birds of America*. The University Society, 1917.

Terres, J. K. *The Audubon Society Encyclopedia of North American Birds*. New York: Alfred A. Knopf, 1980.

Thompson, Lawrance. *Robert Frost: The Early Years, 1874–1915*. New York: Holt, Rinehart and Winston, 1966.

Walsh, John Evangelist. *One Day at Kitty Hawk: The Untold Story of the Wright Brothers and the Airplane*. New York: Thomas Y. Crowell Co., 1975.

Watson, Lyall. *Heaven's Breath: A Natural History of the Wind*. London: Hodder & Stoughton Ltd., 1984.

Wescott, Lynanne, and Paula Degan. *Wind and Sand: The Story of the Wright Brothers at Kitty Hawk*. New York: Harry N. Abrams, 1983.

Chapter Eight: *Convergences*

"Anything but oil." *The Economist*, May 26, 1990, 23–24.

Dolan, R., and H. Lins. "The Outer Banks of North Carolina," *U.S. Geological Survey*, Professional Paper 1177-B, 1986.

Flint, R. F., and B. J. Skinner. *Physical Geology*. New York: John Wiley & Sons, 1974.

Gerding, Mildred, ed. *Fundamentals of Petroleum*. 3rd ed. Austin: University of Texas Petroleum Extension Service, 1986.

Hay, Keith D. *Fish and Offshore Oil Development*. Washington: American Petroleum Institute.

Holing, Dwight. *Coastal Alert: Ecosystems, Energy, and Offshore Oil Drilling*. Washington: Island Press, 1990.

Inman, Douglas J., and Robert Dolan. "The Outer Banks of North Carolina: Budget of Sediment and Inlet Dynamics Along a Migrating Barrier System." *Journal of Coastal Research* 5 (1989): 193–237.

McKibben, Bill. *The End of Nature*. New York: Random House, 1989.

Moser, John J. "Boom or bane: Oil issue is slippery." *Greensboro News & Record*, July 10, 1989: A1. Pt. 2 of a series.

———. "Oil companies seek bite of nature's bounty." *Greensboro News & Record*, July 9, 1989: A1. Pt. 1 of a series.

N.C. Outer Continental Shelf Office. *The Natural Resources Associated with Mobil's Proposed Drill Site*. Proceedings 1989 Marine Expo. Raleigh, N.C., 1989.

Saving Cape Hatteras Lighthouse from the sea: Options and policy implications. National Research Council report. Washington: National Academy Press, 1988.

Schneider, Keith. "Striving for a cleaner way to drill offshore oil." *New York Times*, March 10, 1991: 14A.

Umpleby, S. A. "World Population: Still ahead of schedule." *Science* 237 (1987): 1555–56.

U.S. Congress, Office of Technology Assessment. *Coping With An Oiled Sea: An Analysis of Oil Spill Response Technologies*. Washington: GPO, 1990.

U.S. Department of the Interior, Minerals Management Service. *Leasing Energy Resources on the Outer Continental Shelf*. Washington: GPO, 1987.

U.S. Department of the Interior, Minerals Management Service. *Managing Oil and Gas Operations on the Outer Continental Shelf*. Washington: GPO, 1986.

Index